DATE DUE

Winning Strategies for Power Presentations

Winning Strategies for Power Presentations

Jerry Weissman Delivers Lessons from the World's Best Presenters

Jerry Weissman

Vice President, Publisher: Tim Moore

Associate Publisher and Director of Marketing: Amy Neidlinger

Executive Editor: Jeanne Glasser

Editorial Assistant: Pamela Boland

Operations Specialist: Jodi Kemper

Marketing Manager: Megan Graue

Cover Designer: Chuti Prasertsith

Managing Editor: Kristy Hart

Project Editor: Anne Goebel

Copy Editor: Geneil Breeze

Proofreader: Kathy Ruiz

Indexer: Lisa Stumpf

Senior Compositor: Gloria Schurick

Art Consultant: Nichole Nears

Manufacturing Buyer: Dan Uhrig

Publishing as FT Press

Upper Saddle River, New Jersey 07458

FT Press offers excellent discounts on this book when ordered in quantity for bulk purchases or special sales. For more information, please contact U.S. Corporate and Government Sales, 1-800-382-3419, corpsales@pearsontechgroup.com. For sales outside the U.S., please contact International Sales at

international@pearsoned.com.

Printed in the United States of America

First Printing November 2012

ISBN-10: 0-13-312107-0

ISBN-13: 978-0-13-312107-0

Pearson Education LTD.

Pearson Education Australia PTY, Limited.

Pearson Education Singapore, Pte. Ltd.

Pearson Education Asia, Ltd.

Pearson Education Canada, Ltd.

Pearson Educación de Mexico, S.A. de C.V.

Pearson Education—Japan

Pearson Education Malaysia, Pte. Ltd.

Library of Congress Cataloging-in-Publication Data

Weissman, Jerry.

Winning strategies for power presentations : Jerry Weissman delivers lessons from the world's best presenters / Jerry Weissman.

p. cm.

ISBN 978-0-13-312107-0 (hardback : alk. paper) -- ISBN 0-13-312107-0

1. Business presentations. I. Title.

HF5718.22.W453 2013

To my mentors:
Professors Harry Miles Muheim and Ormond Drake
at New York University

Ray Abel and Mike Wallace at CBS

Evelyn Grippo at Pinnacle Books

Table of Contents

The Masters

In order of appearance:

Marcus Tullius Cicero
Aristotle
Mark Twain
Joshua Foer
Maureen Dowd
Ed Cooke
Amanda Foreman
Sir Arthur Quiller-Couch
Christopher Markus
Stephen McFeely
William Strunk
E.B. White
Stephen King
Jason Gay
Franklin D. Roosevelt
James Collins
Maryanne Wolf
Guy Kawasaki
Federico Fellini
Casey Schwartz
Woody Allen
Mick LaSalle
Stephen R. Covey
Terry Teachout
Manohla Dargis
A.O. Scott
Clive Thompson
Michel de Montaigne
Henry James
Tim Carmody
Nicholson Baker
Sir Winston Churchill
Mardy Grothe
Erin McKean
Leroy "Satchel" Paige
Humphrey Bogart
Julius J. Epstein
Philip G. Epstein
Howard Koch
William Schneider
Steve Kelley
Mike Lee
John Doerr
Vinod Khosla
Walter S. Mossberg
Joe Dator
Sherry Turkle
James W. Pennebaker
Diana Tamir
Jason Mitchell
Theodore Leavitt
Bruce Eric Kaplan
Geoff Dyer
John Irving
William Safire
Jhumpa Lahiri
David Letterman
Deepak Chopra
William Shakespeare
William Gillette
Ronald Reagan
Barack Obama
Philip Delves Boughton
Christopher M. Barlow
Paul Katz
Jon Stewart
Akira Kurosawa
Stephen Prince
Matt Zoller Seitz
Lucy Kellaway
Michael Baldwin
Garr Reynolds
Stephen M. Kosslyn
Hans Rosling
Deborah Landau
Stephen Sondheim
Paul Simon
Dizzy Gillespie
Houston Person
Dana Goodyear
W. Timothy Gallwey
Joel Stein
Taylor Mali
Matt Ridley
Irving Berlin
Tony Perrottet
Bruce Iliff
Rolf Dobelli
Frank Sinatra
John F. Kennedy
Jon Meacham
Johnny Carson
Dave Wiegand
Dorothy Rabinowitz
Adam Lashinsky
Frank Partnoy
Scott Adams
Bill Clinton
Kevin O'Connor
Paul Maritz
Adam Bryant
George Bernard Shaw
Peter Funt
Charlie Rose
Barbara Walters

Introduction

Natural and Universal

There is nothing new under the sun.

—Ecclesiastes 1:9

For businesspeople, presentations are an unnatural act.

Presenters are not performers, nor are they graphic designers, nor do they have an abundance of time, and—what is most unnatural of all—whenever they have to deliver a mission critical pitch, their own elevated stress diminishes their effectiveness.

As a result, most business presentations devolve into a mind-numbing scenario in which a nervous person stands in front of a room giving a *verbatim* recitation of a disjointed set of begged, borrowed, or stolen slides to a bored audience for far too long.

In my role as a presentation coach, I sought to end this vicious cycle and found it in the commonality with other communication modes. Presentations are not unique public speaking situations practiced by a privileged few on special stressful occasions; presentations have the same goals and dynamics as meetings, conversations, telephone calls, job interviews, and interpersonal communications. They all aspire to convey ideas between two separate people or groups of people, to ensure that both parties connect, and to ensure that the "co-" in communication is achieved.

Presentations also have the same goals and dynamics of broader communication modes such as literature, cinema, media, and politics. Many of the expert practitioners in these fields have shared their secrets in public, so I have tapped into their advice and adapted them into a set of best practices that you can use in your presentations. In these pages you'll find the wisdom of Mark Twain, Woody Allen, Johnny Carson, Ronald Reagan, and many other leaders in their field of communication, with special mention to Scott Adams, the creator of *Dilbert*, the wonderful comic that strip satirizes *dysfunctional* communications in business.

The place of honor, however, goes to Marcus Tullius Cicero, the great Roman statesman and orator, whose highly functional advice forms bookends in the first and culminating chapters of this volume, as well as here in the Introduction, where his words, written in 55 BC, support the natural approach:

> *The special province of the orator is, as I have already said more than once, to express himself in a style at once impressive and artistic and conformable with the thought and feeling of human nature.*[1]

This universal vision was reinforced in 2012 when Cheers Publishing in Beijing translated my first three books on presentation skills into Chinese and released them as a trilogy. Originally, I wrote *Presenting to Win: The Art of Telling Your Story, The Power Presenter: Technique Style and Strategy*, and *In the Line of Fire: How to Handle Tough Questions* as individual books rather than as an omnibus so that I could provide readers with a thorough methodology for each of the essential elements of every presentation:

- How to develop a clear and logical story
- How to design simple and effective graphics
- How to speak with confidence and authority
- How to handle challenging questions

Seeing the three books together and in Chinese (even though I did not understand the Chinese characters) validated my view that the essential elements of any presentation have the same roots and—except for PowerPoint—have existed since Cicero's time in ancient

Rome, and even earlier, in Aristotle's time in ancient Greece. The principles established by those classical philosophers are still applicable today. I have been using modern versions of them in the public and private programs of my coaching practice in Silicon Valley for almost a quarter of a century, and for a decade before that at WCBS-TV in New York in my role as a producer and director of public affairs programs.

To share these timeless and borderless practices with you, I've crafted them as individual lessons in succinct, bite-sized, chapters. I used the same approach in my previous book, *Presentations in Action*, as well as in blogs posted on the Forbes and Harvard Business Review websites, and on indezine.com, a dedicated PowerPoint site, where some of these new lessons have previously appeared.

Beyond presentations, you'll also find advice on how to handle special speaking situations such as large audience formats, panel discussions, product demonstrations, interviewing, scripted speeches, and voice and speech quality. And for those of you fortunate enough to reach the top of the business mountain, I've also included ten best practices for my specialty, the Initial Public Offering road show.

I've had the privilege of coaching the IPO road shows of nearly 600 companies, among them Cisco, Intuit, Yahoo!, eBay, Netflix, and Dolby Laboratories. For each of them, I used the same techniques as I did with another 600 companies, coaching them to develop presentations to raise private financing, sell products, form partnerships, and gain approval for internal projects—further substantiation of the universality of this methodology.

At the foundation of all these applications is the larger message that you can access and employ the same best practices that have proven successful over time and across diverse geographies, cultures, and media, to help you to become a Power Presenter.

You have my very best wishes for success.

Section I

Content: The Art of Telling Your Story

1

Mark Twain's Fingernails

How to Remember What to Say

A subject close to the pounding hearts and racing minds of every public speaker or presenter is how to remember what to say. Speakers and presenters routinely rely on a number of devices from low-end 3-by-5 index cards to expensive high-end teleprompters—to aid their memories.

Joshua Foer, the author of the bestselling *Moonwalking with Einstein: The Art and Science of Remembering Everything*, offers an even higher end but lower cost technique: visual imagery, or associating a diverse list of subjects with a series of related objects. Mr. Foer's physical take on mnemonics is only the latest variation of a method called "loci," (from the Latin word for "places") which, according to a *Wall Street Journal* article,[1] has roots that go back to our cave dwelling ancestors because it helped "humans remember which trails through the woods lead back home."

The loci method was used in ancient Rome by Marcus Tullius Cicero, the first century philosopher, statesman, and orator. When Cicero and his contemporaries delivered their lengthy speeches in the Roman Forum, they spoke without notes because paper had yet to be invented. So the speakers used the marble columns of the Forum as memory triggers. Each column represented a single subject and its related ideas. As the tour guides at the ruins of the Forum tell it today, the orators delivered their speeches, striding from column to column and subject to subject, using the columns as visual prompts to remind them of a group of related ideas.

Over the years, this technique has morphed into the popular "Roman Room" memory method, in which physical objects inside a room serve the same associative purpose as the open air columns of the ancient Roman Forum.

Maureen Dowd of the *New York Times*, inspired by Mr. Foer's book, found two other writers with intriguing memory aids:

- Mark Twain, who "once wrote the first letter of topics that he wanted to cover in a lecture on his fingernails."
- England's Ed Cooke, the author of *Remember, Remember* and a Roman Room devotee who recommends, "If you have a list to remember, you put the items in a path throughout a familiar place, like your childhood home."[2]

Mr. Cooke, who is also the co-founder of Memrise,[3] a website focused on memory, related the technique directly to presentations. In a 2008 article in London's *The Guardian*, he wrote:

> *Begin by reducing your talk to, let's say, 20 bullet-points.... Write out your points in order. Now find an image that captures each point. To remember that the pound is losing ground on the dollar, you could imagine George Bush beating up Gordon Brown with a wad of dollar bills. If you wish to remember that 90% of women are at a disadvantage in the workplace, you might imagine a 90-year-old woman carrying a heavy weight. Then arrange your images on a route around a familiar space. So the Bush-Brown scenario could go in your bathroom sink, the granny could go in your shower, and the next 18 images could be arranged sequentially in a route around your home.*[4]

In my coaching version of Mr. Cooke's advice, I go back to Cicero and recommend that speakers and presenters cluster the diverse components of their pitches into a few conceptual Roman columns, or main themes; and then to represent those ideas in simple PowerPoint slides designed under the Less is More principle. The memory prompt then comes from a specific image rather than from an imaginary physical layout.

Financial executives, with their usual attention to detail and concern about forward-looking statements, often prepare their presentations as complete text on paper or on slides, and then they read or try to memorize the words. Those approaches force the presenter to stay connected to the text and disconnected from the audience.

One public company's chief financial officer showed up for his coaching session with his presentation written out in full sentences. I asked him to reduce each sentence to a four-word bullet and to speak from that. He did and it flowed. Then I asked him to reduce each four-word bullet to one word and to speak from that. He did and it flowed. Then I asked him to speak without any text. He did and it flowed. We then put the four-word bullets on the slides and he delivered his pitch directly to the audience and it flowed.

Of course, you can always skip the PowerPoint slides and, like Mark Twain, write the first letter of each of your subjects on your fingernails or, like Sarah Palin, write notes on your palm. Then again, you can default to those old standby 3-by-5 index cards. But every time you glance down, you will not only disconnect from your audience, you will also appear to be unsure about what to say and diminish your credibility.

Better to go with Cicero's columns and PowerPoint.

2

Kill Your Darlings

A Lesson from Professional Writers

Historian Amanda Foreman, the author of the bestselling *Georgiana, Duchess of Devonshire*, followed it a decade later with *A World on Fire: Britain's Crucial Role in the American Civil War*. In describing her creative process for the *Wall Street Journal's* "Word Craft" column, she provided a valuable lesson for presenters:

> *The fruit of my 11 years of research meant that I had more than 400 characters scattered over four regions.... This vast mass of material was so unwieldy that I could hardly work my way through the first day of the conflict, let alone all four years.*[1]

While few presenters spend 11 years developing their stories about their businesses, they, like Ms. Foreman, have a vast mass of unwieldy material that they have to communicate to various audiences. Unfortunately, most presenters then proceed to deliver that mass to their audiences in full, inflicting the dreaded effect known as MEGO or "My Eyes Glaze Over."

Although Ms. Foreman is a respected scholar with a doctorate in history from Oxford University, she has storytelling in her DNA.[2] Her father was Carl Foreman, an Oscar-winning screenwriter who wrote the classic film *The Bridge on the River Kwai*. At the end of her research, Ms. Foreman realized that, even for a story as immense and complex as the Civil War, she had too much information for both writer and reader to process. Her solution:

I plotted the time lines of my 400 characters and identified and discarded people who, no matter how interesting their stories, had no connection to anyone else in the book. This winnowed my cast down to 197 characters, all bound to one another by acquaintance or one degree of separation. ✲

Ms. Foreman was tapping into a practice—well-known among professional writers—called, "kill your darlings." In fact, a community of writers in Atlanta has adopted that name for its website.[3] The phrase is often attributed to Nobel Prize novelist William Faulkner, but it was actually coined by Sir Arthur Quiller-Couch, a British writer and critic who, in his 1916 publication *On the Art of Writing*, said:

Whenever you feel an impulse to perpetrate a piece of exceptionally fine writing, obey it—whole-heartedly—and delete it before sending your manuscript to press. Murder your darlings.[4]

The sentiment was echoed by Christopher Markus and Stephen McFeely, the screenwriters of *Captain America*, the Hollywood action film based on the 70-year-old comic strip character. In another *Wall Street Journal* "Word Craft" article, the team wrote:

Adapting an existing work for film is usually a process of reduction. Whether it's a novel or a short story, a true-crime tale or 70 years' worth of comic books, the first job is distillation. If this means losing someone's favorite character, so be it. The simple fact is that we can't put everything on the screen. Darlings must die.[5]

The phrase rings true because writers, who labor over their ideas and words like expectant mothers, invariably fall in love with their offspring and are reluctant to find fault, and even more reluctant to part with them. In the same manner, presenters who live, breathe, walk, talk, and dream about their businesses, want to share every last detail with their prospective audiences. But audiences do not share their interest, and so presenters, like writers, must kill their darlings.

✲ Amazon lists Ms. Foreman's book at 1,008 pages. Imagine how many more pages it would have run had she not killed those 203 characters.

In presentations, the process begins by assembling all your story elements. A chef prepares for a meal by gathering all the ingredients, seasonings, and utensils, but doesn't use every last one of them. Once you have assembled all your presentation ingredients, assess every item for its relevance and importance to your audience—*not* to you. Your audience cannot possibly know your subject as well as you do, and so they do not need to know all that you do. Tell them the time, not how to build a clock.

Delete, discard, omit, slice, dice, or whatever surgical method you choose to eliminate excess baggage. Be merciless. Retain only what your audience needs to know.

Once you have made that first cut, make another pass, and then another. Each time you do, you see your draft with fresh eyes and find another candidate for your scalpel. Follow the advice of the classic Strunk and White's *The Elements of Style*: "It is always a good idea to reread your writing later and ruthlessly delete the excess."[6]

Bestselling horror novelist Stephen King—who knows a thing or two about ruthless killing—follows a similar practice. In his 2000 book *On Writing*, he shared a note his editor once sent to him: "You need to revise for length. Formula: 2nd Draft = 1st Draft – 10%."[7]

Deal with your vast mass of unwieldy material in your preparation, *not* in your presentation; behind the scenes, *not* in front of the room. A gentler way of saying "kill your darlings" is "when in doubt, leave it out."

3

How Long Should a Presentation Last?

I can well understand the Honourable Member's wishing to speak on. He needs the practice badly.

—Sir Winston Churchill[1]

Be Brief and Concise

Every presenter is painfully aware of the short attention span of modern audiences. Jason Gay, who writes for the *Wall Street Journal,* captured the essence of the phenomenon perfectly:

> *You are busy, busy, busy! This is presumed in 2012. You are too hurried to pay cash for your coffee—oh, it's so arduous, removing the bills from the pocket, unfolding them, handing them to the cashier, waiting for change, which can take up...to...nine...agonizing...seconds. So you pay with an app on your smartphone—blip, done.*[2]

But short attention spans pre-date 2012; they go all the way back to the twentieth century.

President Franklin D. Roosevelt, one of America's, if not the one of the world's, most accomplished orators answered the question in the title of this chapter with his famous advice, "Be sincere, be brief, and be seated."[3]

This wise counsel is sadly all too often ignored in most presentations where brevity is a rarity. FDR was referring to the length of a speech, but brevity is also important in regard to the amount of detail in a presentation.

James Collins, author of the novel *Beginner's Greek,* addressed the subject in a *New York Times* Book Review essay titled "The Plot Escapes Me." Mr. Collins described how much he likes to read books, but that once he finishes, "I remember nothing about the book's actual contents...all I associate with them is an atmosphere and a stray image or two." He went on to note that he is not alone, "most people cannot recall the title or author or even the existence of a book they read a month ago, much less its contents."[4]

Curious about the phenomenon, Mr. Collins discussed it with Maryanne Wolf, a professor at Tufts University and Director of Center for Reading and Language Research. Professor Wolf validated the experience. "There is a difference," she said, "between immediate recall of facts and an ability to recall a gestalt of knowledge. We can't retrieve the specifics, but to adapt a phrase of William James's, there is a wraith of memory."

As important as controlling the amount of content is in written text, it is even more important in live presentations. Readers of text—whether in print or electronic form—can always go backward in the text to clarify details, but audiences for presentations—who receive the content in real time—do not have that option. If they lose track, they will either interrupt the presenter or tune out; at which point the presenter loses too.

The lesson for you is to be concise. Be mindful of not only the overall length of your presentation, but also the amount of detail you include and—just as important—how you organize those details. Craft your presentation with just three to five high level themes, and be sure that any information you include clearly relates to those themes. Then, as you present, keep referencing the themes and tying the details back to them. Your audience may forget the details after your presentation, but they will "recall a gestalt" of your story.

Or as Guy Kawasaki, that master of well-turned phrases, in his own variation of FDR's advice, wrote in his book *Enchantment*, "Make it short simple and swallowable."[5]

4

Follow the Money

"So...?"

The infamous Watergate affair—captured so effectively in *All the President's Men*, the 1974 bestselling book by *Washington Post* reporters Bob Woodward and Carl Bernstein, and the 1976 film version starring Robert Redford and Dustin Hoffman as the reporters—produced many memorable phrases that have worked their way into our culture. One of the most enduring was the line spoken by the clandestine informer, "Deep Throat," who advised Mr. Woodward and Mr. Bernstein to "follow the money."

Presenters can follow the same advice in their pitches. However, in their eagerness to load up their stories with features, and their even greater eagerness to get back to their seats, presenters often neglect to provide a benefit to their audiences. In the process, they lose the money trail.

In *Presenting to Win*, I provided an illustration that bears referencing here. A start-up company seeking financing asked me to coach their pitch to a venture capital firm. During the presentation, the CEO presented a slide listing the benefits of his company's product line. After he discussed each of them, he summarized by saying, "You can see that our products provide a rich set of benefits to our customers."

I said, "So...?"

The CEO smiled and added, "These benefits bring our company repeat business, repeat business brings us recurring revenues, and recurring revenues grow shareholder value."

Those were benefits for the venture capitalists. That's following the money.

Another illustration: Karen Wespi, a sales manager for Maxim Integrated, a company that designs and markets analog and mixed signal semiconductors, participated in a program to develop a pitch to sell Maxim's products to a large computer manufacturer. During her presentation, Ms. Wespi presented a slide listing the many features of the company's product line, discussed each of them in detail, and then summarized by saying, "So you can see that Maxim's products provide the best technology with the most bang for the buck."

I said, "So...?"

Ms. Wespi then added, "Maxim's integrated solutions will enable you to add greater performance to your computers, bring them to market faster, and give you an advantage over your competitors."

She gave the benefits for the computer manufacturer. That's following the money.

In developing your presentation, repeatedly ask yourself the "So...?" question. Put yourself in place of the start-up CEO or Ms. Wespi; or put your audience in place of the venture capital firm or computer manufacturers, or *any* audience. When you go on to provide the answer to that question, you'll take a step forward on the money trail.

Deep Throat had it right: Follow the money.

5

Fellini on Creativity

Consider All the Possibilities—Before You Present

Federico Fellini, the legendary Italian film director noted for his imaginative cinematic works, had some very explicit ideas about the creative process. In *Fellini on Fellini*, his book about his art, he described how he generates ideas:

> *I hate logical plans....Myself, I should find it false and dangerous to start from some clear, well-defined complete idea and then put it into practice....The child is in darkness at the moment he is formed in the mother's womb.*[1]

Crafting a presentation is a creative process. In what has become standard operating procedure in business, most presenters reverse Mr. Fellini's approach. They start with a "clear, well-defined idea,"—usually in the form of a set of company slides—and then "put it into practice" by standing up to present. The result is the predictable data dump.

That's because this approach reverses the natural functions of the human mind, known among psychologists as "divergent" and "convergent" thinking. Neuroscientist Casey Schwartz described the difference in a blog about creativity on *The Daily Beast* website:

> *Divergent thinking is the ability to generate spontaneous, often unexpected ideas or solutions ... Convergent thinking, on the other hand, is understood as divergent thinking's opposite: the kind of thought process that allows you to narrow down your options.*[2]

When presenters begin their creative process with slides, they are narrowing their options for ideas. The solution is to do your convergent thinking *after* your divergent thinking; to let your mind to do what it is going to do anyway: generate ideas randomly—and then capture them in brainstorming.

Before you even consider your slides, consider all the *ideas* you want to discuss, but treat them as words, not images. If you start with your slides, you front load your mind with everything from the color or size of the font to a pre-existing sequence. Instead, start with your ideas and write them on paper, or on a computer screen, a whiteboard, or Post-it Notes. Then look at *all* of the ideas objectively and decide which ones you need and—more important—which ones you don't need.

Do the data dump in your preparation not in your presentation. Do your divergent thinking *before* your convergent thinking.

Get creative.

6

How Woody Allen Creates

First Things First, Last Things Last

Woody Allen, a virtual one-man movie studio, having written more than 60 films during his long and illustrious career, has his own version of the creative process used by his colleague, Federico Fellini, the Italian director about whom you read in the previous chapter: free-form thinking.

Unfortunately, most presenters, in their rush to prepare their next pitch, begin by shuffling existing slides, and often at the last minute. They do this because, as results-driven people, they seek to impose structure at the outset. But every human mind, whether artistic or business, generates ideas randomly, so an essential part of the creative process—and developing a presentation is a creative process—is to incorporate the randomness. Artists understand this fact of life and go with the flow.

Mr. Allen revealed his creative process in a biographical documentary on the *American Masters* series on PBS. In a scene shot in his apartment, Mr. Allen reached into a nightstand drawer, took out a large stack of cluttered papers and said, "This is my collection. This is how I start. It's all kind of scraps and things that are written on hotel things. I'll ponder these things." Then, as he tossed the papers onto his bed, he added, "I'll dump them here like this...I go through this all the time, every time I start a project. And I sit here like this...and I look at one...like that...and then...."[1]

For your brainstorming, as your version of Mr. Allen's hotel scraps, you can use 3-by-5 index cards, a whiteboard, Post-it Notes or one of the many software products on the market, among them Inspiration, MindManager,[2] and Microsoft's Visio.[3] Whichever vehicle you choose, consider any and all ideas—but resist your results-driven instinct to impose structure during your free flow. If you impose structure too soon, you impose censorship and could lose a fresh idea. Save the structuring for *after* the brainstorming is done.

Here, too, we find a lesson in the methodology of Woody Allen and Federico Fellini. Each of them is noted for his creativity in post-production, the period *after* the writing and the shooting, when the director assembles and structures the film, However, Mr. Allen's assemblage of his 2012 production *To Rome with Love* did not impress A. O. Scott, the *New York Times* critic. Mr. Scott called the film a "genial tangle of stories (which Mr. Allen seems to have unpacked from a steamer trunk full of notes and sketches)."[4] Apparently, Mr. Scott saw the documentary, too.

Mr. Fellini took post-production to a new level of creativity. He cast actors who looked best for the filming and other actors whose voices sounded best for the sound track and overdubbed them during the editing process.

Let your mind do what it's going to do anyway—during your brainstorming—then do your structuring afterwards. Use the right tool for the right job and in the right sequence.

Follow Woody Allen's advice, "It's not rocket science, this is not quantum physics. If you're the writer of the story, you know what you want your audience to see because you've written it. It's just storytelling and you tell it."[5]

7

What's Your Point?

Leave Pointlessness to Woody Allen

Existentialism is defined by Wordnik, the online dictionary, as "A philosophy that emphasizes the uniqueness and isolation of the individual experience in a hostile or indifferent universe, and regards human existence as unexplainable."[1]

Woody Allen, who has made frequent references to the philosophy in his films—name-dropping its advocates, Kierkegaard, Nietzsche, and Sartre—is reported to have once said, "I took a test in Existentialism. I left all the answers blank and got 100."[2]

In his 40th film as a director, *You Will Meet a Tall Dark Stranger*, Mr. Allen turned to the subject again, if not in philosophical quotes, in theme. As a matter of fact, the only—and most pertinent—quote he used is by William Shakespeare. At the beginning of the film, an off-screen narrator speaks these famous lines from *Macbeth*: "A tale told by an idiot, full of sound and fury, signifying nothing."

The tale that Mr. Allen then goes on to tell is about two couples whose marriages break up because each of the members strays in search of a better partner. But in the end, all four searchers wind up in circumstances worse than they had left; their failed searches pointless.

In his review of the film, Mick LaSalle, the movie critic of the *San Francisco Chronicle*, wrote:

> *Allen once again contemplates the pointlessness of existence, but this time he has an additional idea, one that has an effect on his movie's story and structure.... His goal is not to make you walk out thinking, "Ahh, yes, perhaps there is no moral order to the universe. Very interesting." His goal is rather to make you walk out thinking, "Huh? What was the point of that?"*[3]

"What was the point of that?" and its companion phrase, the teenagers' frequently-uttered, "And your point is...?" are the very last words you want your audience to walk out thinking at the end of your presentation. How many times have *you* have been in the audience to someone else's presentation and muttered those disdainful words?

You can avoid that reaction in your audience by following the advice of author Stephen R. Covey, whose bestselling *The 7 Habits of Highly Effective People* identifies Habit 2 as "Begin with the End in Mind."[4] Start the development of your presentation with the last sentence; then build up to it with strong, powerful ideas and words.

Woody Allen can get away with pointlessness because he goes for the laughs; you are going for the gold. Make your presentations full of sound and fury, signifying *everything* that you want your audience to do.

8

Spoiler Alert

What's Your Point?

Should drama or film critics reveal the ending of a play or a movie in their reviews—especially when the ending is a surprise? Should a review of *Citizen Kane* reveal the identity of "Rosebud"?

Terry Teachout, the *Wall Street Journal's* drama critic, thinks not. In his review of the hit play at New York Lincoln Center, *War Horse*, he vowed silence twice, citing "critical etiquette" and "the drama critic's code."[1]

Manohla Dargis and A. O. Scott, the co-chief film critics of the *New York Times*, disagree, and their opinions about revelations provide an important lesson for presenters. Mr. Scott has no qualms about letting the cat out of the bag:

> *Anna Karenina dies at the end. Madame Bovary too. Also Hamlet and just about everyone else in* Hamlet.

Nor does Ms. Dargis:

> *Seriously, if you don't want to know what happens in a film, book, play or television show, you shouldn't read the reviews until after you've watched or read the work yourself. (I rarely do.) Because no matter how delicately a critic tiptoes around the object, she invariably reveals something that someone resents, whether it's a bit of plot, a line of dialogue or...a shameless finale.*[2]

Other critics hedge their bets by adding a "spoiler alert" to their reviews, advising readers that vital plot information is to follow.

Surprise revelations are all well and good for plays and films because the suspense keeps theater audiences glued to their seats and movie audiences buying tubs of popcorn, but suspense is not good for presentation audiences. They have neither the time nor the inclination for such tactics. The last thing you want your audiences to think is, "What's your point?"

In the previous chapter, you read about how important it is to state your point clearly in presentations. Here we raise the bar: State your objective at the *very beginning* of your presentation—within the first 90 seconds. With all due respect to Mr. Teachout, you must be as open about your point at the start of your pitch as Ms. Dargis and Mr. Scott are about endings. A playwright or a filmmaker can wait until the very last scene to reveal that the butler did it; you do not have that luxury.

You never get a second chance to make a first impression.

Get to the point!

9

The Cyrano Parable

The Story You Tell Versus the Slides You Show

Ever since its debut in Paris in 1897, *Cyrano de Bergerac* has held enormous attraction for audiences and performers. The original French actor, for whom playwright Edmond Rostand created the part, performed it more than 400 times. In the twentieth century, the story went on to be produced as a film six times—one of which earned an Oscar for José Ferrer in 1950, and another that starred Steve Martin in 1987—and as an opera five times.

While the story is based on a real seventeenth century poet and swordsman, it was Mr. Rostand's nineteenth century interpretation that created its enduring appeal—and serves as a parable for presenters. Briefly stated, *Cyrano* is the story of a man who was considered ugly because of his exceptionally large nose, but who more than compensated for his looks with a rare gift for language. In a tale of romance by proxy, Cyrano helps a handsome but inarticulate man win over a beautiful woman by writing love letters for him and by speaking for him at a masquerade ball, almost as a ventriloquist.

The Cyrano Parable is a testament to the power of substance over style and the vital importance of the story. It is also an analog for the primacy of the presenter's narrative over the slide show. This hierarchy is supported and promoted by a cottage industry of presentation consultants, authors, coaches, designers, websites, and organizations. I heard many of them reinforce this approach as a virtual mantra at

Rick Altman's Presentation Summit, an annual industry conference for graphics professionals.[1]

Nevertheless, their clients and my clients—businesspeople all over the world—continue to ignore the advice and follow the opposite approach of having their slides tell their stories. While this practice produces a never-ending deal flow for consultants and coaches, it rarely produces successful presentations for presenters. Heed the advice of presentation professionals; keep your slides simple and tell your own story.

PowerPoint is not a ventriloquist, and you are not a dummy.

10

"Does that make sense?"

...And Other Meaningless Words

Language is alive, a dynamically evolving and changing entity.

One of the newest expressions to gain momentum in American speech is, "Does that make sense?" The phrase is most often used by a speaker in the middle of a conversation—or a presenter in the middle of a presentation—to check whether the listener or audience member has understood or appreciated what the speaker has just said. Unfortunately, the expression has three negative implications:

- Uncertainty on the part of the speaker about the accuracy or credibility of the content
- Doubt about the ability of the audience to comprehend or appreciate the content
- The word, "sense," which, by its interrogative formation, implies that the content in question is of dubious sense, or senseless, or complete nonsense

"Does that make sense?" has become so pervasive it has taken its place among other filler words such as "I'm like..." and "I mean..." Most speakers are unaware they are saying it, and most audiences don't bother to think of its implications. The phrase has attained the frequency—and meaninglessness of:

- "You know..." as if to be sure the listener is paying attention
- "Like I said..." as if to say that the listener didn't understand

- "Again..." as if to say that the listener didn't get it the first time
- "To be honest..." as if to say the speaker was not truthful earlier

Every responsible speaker or presenter, in their well-intentioned effort to satisfy their audience, has every right to check whether their material is getting through and whether their audience is satisfied. But instead of casting negativity on the clarity of the content or the intelligence of the audience, ask instead, "Do you have any questions?"

This neutral question will produce either further clarifying questions or acceptance that the content is clear—in both cases, the closure of a complete send-receive loop between the presenter and the audience, the endgame of all communication.

11

Meaningful Words

Words That Inspire Confidence

In the previous chapter, I recommended that you eliminate meaningless words that are condescending, insulting, or self-deprecating; but there is another group of meaningless words that require replacement. The list begins with:

- Believe
- Think
- Feel

Attorneys have long cautioned officers and employees of corporations to avoid forward-looking statements. The financial scandals of the past decade have made those attorneys even more diligent about language. As a result, corporate presenters now fill their pitches with sentences formed in the conditional mood. Phrases containing "we believe..." "we think..." and "we feel..." pervade presentation narratives to such a degree that they spill over into sentences where caution is unnecessary. More to the point, the spillage weakens what should otherwise be assertive language, as in the following sentence:

With this large opportunity and our superior technology, I think you'll see that our company is well-positioned for growth.

The words "I think" introduce doubt, even if only subliminally, in the minds of your audience. As a presenter attempting to persuade an audience, your job is to provide them with as much certainty as you can. The way to get from doubt to certainty is to switch from the conditional to the declarative mood by eliminating the offending words:

With this large opportunity and our superior technology, you'll see that our company is well-positioned for growth.

This simple nip and tuck strengthens the impact of the entire sentence.

This is not to say that, when the outcome is uncertain, you should make forward-looking statements or forecasts. That's risky business. In such cases, you must use the conditional mood, but instead of the weak words "think," "believe," and "feel," replace them with these stronger options:

- We're confident...
- We're convinced...
- We're optimistic...
- We expect...

Hear the difference? Now look at the earlier sentence made stronger:

With this large opportunity and our superior technology, you'll see that our company is well-positioned for growth, and we're confident that growth will translate into significant revenues.

From the sublime of persuasive words to the banal of airline travel, think of the announcement you typically hear on the public address system when your flight touches down at your destination:

I'd like to be the first to welcome you to San Francisco.

Sound familiar? It's boilerplate; not just in airline travel, but also in political speeches, college lectures, church sermons, award ceremonies, acceptance speeches, wedding toasts—the list is endless. In business presentations the sentence sounds vague and indefinite.

Besides, if you'd like to do it, why not just go ahead and do it?

Welcome to San Francisco!

Meaningful words stated assertively in the declarative mood, are more likely to beget meaningful actions.

12

Writer's Block

How to Break Through

Writer's block is the proverbial stuff of legend and literature. A variation on the theme is *Limitless,* a Hollywood film starring Bradley Cooper and Robert De Niro. In it, Mr. Cooper plays a down-and-out writer who beats his severe case of writer's block with a new drug that not only jump-starts his creative output, but gives him many other advanced mental capabilities. Of course, the story is fictional—A.O. Scott's review of the film in the *New York Times* called it, "an energetic, enjoyably preposterous compound... a paranoid thriller blended with pseudo-neuro-science fiction and catalyzed by a jolting dose of satire"[1]—but the situation is very real: Writers do run dry.

Mr. Scott went on to list the many real attempts tortured writers have made to get past their paralysis: "Sharpen 10 pencils. Eat a sandwich. Pretend that the first chapter of your long-overdue opus is a casual letter to your grandmother. Weep quietly. Have another drink."

However, creative block is not limited to aspiring and professional writers who are trying to craft their next article on deadline or write The Great American Novel. Presenters, too, are frequently faced with having to crank out their next great pitch. Their bar is not as high as that of a solitary writer staring at a blank computer screen or a yellow legal pad. Presenters belong to a team—a business unit in a large company or a small start-up—and so they have access to their colleagues' slide shows.

Therein lies the problem: Businesspeople consider their presentations to exist primarily in PowerPoint. Many companies amass a large, searchable database of slides for anyone in the organization to access. Enter "corporate strategy" and dozens of slides containing those words download in an instant. The problems is then compounded when a presenter picks out what he or she thinks are the appropriate slides and then assembles them in an order that is meaningful to him or her—but *only* to him or her. The resultant aggregation is what is known as a "Frankendeck."

It gets worse. Having to rely on a set of disparate slides created by others, the presenter reads the slides verbatim to the audience. The inevitable result is a train wreck.

The problem with this method of preparation is that it starts in middle of the creative process and then jumps to the end, skipping several important steps along the way.

A simple solution is to begin at the end instead. Do what author Stephen R. Covey recommends. As you read in Chapter 7, "What's Your Point?," begin with the end in mind, which in a presentation, is the goal or objective of your pitch.

By beginning with the objective and *not* with the slides, the entire story has an overarching focus. Then, still working without slides, follow these next steps:

- Analyze your audience and how they will react to your objective.
- Brainstorm ideas that support your objective and address your audience's needs.
- Identify the key ideas and discard the irrelevant ones.
- Organize the key ideas into a logical flow.

Only then are you ready to design slides that serve their sole purpose: to illustrate the key ideas.

This step-by-step process is a prescription not for a magical drug but a solution that will enable you to realize your own creative process—and create winning presentations.

13

Writer's Block II

Easier Said Than Done

The step-by-step process to get past the proverbial blank page of the previous chapter is as applicable to presenters as it is to writers. Another method to break through the mental barricade is to just start talking.

Writers have long known that speaking aloud what they have written in silence helps them to shape their ideas. In a *Wired Magazine* article on voice recognition, Clive Thompson tells of sixteenth-century French essayist Michel de Montaigne and nineteenth century American writer Henry James, both of whom wrote by dictating their work to their secretaries. Moving to the present, Mr. Thompson cites the example of writer and critic Tim Carmody who "found himself staring at an empty page, not knowing where to begin. He had no problem talking to friends about his ideas, so Carmody booted up Dragon (voice recognition software from Nuance), talked aloud for hours, and got past the block."[1]

Mr. Carmody was experiencing the starting point of a spectrum of benefits that comes from combining the written words with the spoken. Speaking aloud also provides perspective during the homestretch of the creative process—in the reviewing and polishing steps. Many professional writers read their work aloud to themselves (rather than to their secretaries as Messrs. Montaigne and James did).

Giving sound to what had been a silent process puts writers in the role of their readers. This gives writers an objective view of their content. Bestselling author Nicholson Baker calls his version of this process "speak-typing," in which he dictates to himself and types as he speaks. In an interview with the *New York Times* about his book *House of Holes,* Mr. Baker explained that "the words come out differently. The sentences come out simpler, and there's less of a temptation to go back and add more foliage. I'm trying for a simpler kind of storytelling."[2]

Presentations are all about speaking aloud, and preparing for them should involve talking too. As a coach, I recommend that presenters rehearse their presentations by displaying their PowerPoint slides in the Slide Sorter view (think of this as the Storyboard) and then running through their narrative aloud, assuming the role of their audience.

But giving voice to ideas also helps that challenging front end of the creative process. Just as Mr. Carmody did, you can jump-start your own creative process by speaking your presentation aloud and recording it using Dragon software or the voice record function on your smartphone. Play back the recording afterwards to shape or reshape your ideas and words, but the key to breaking the logjam is to start talking.

Writer's block occurs because the prospect of starting from scratch is daunting. Even if a writer has a clear idea of a new story—or a presenter has a clear idea of a new presentation—the prospect of choosing which of all the available ideas to include or how much detail to provide, overloads the writer's mind. However, writers and presenters alike, having lived with their subject matter, know it intimately and have no difficulty chatting about it. Extend that facility into having a private chat with your recording device. You'll find the process liberating and productive.

Mr. Thompson's article tells us how much Mr. Montaigne valued the process: "'The things I say,' Mr. Montaigne dictated, 'are better than those I write.'"[3]

14

Never Say "Never"

Well, Almost Never

Sir Winston Churchill, the great British Prime Minister, prolific author, and distinguished orator who addressed some of the most august assemblies in the world, once delivered a speech to the boys at Harrow School in Britain:

> *Never, never, in nothing great or small, large or petty, never give in except to convictions of honour and good sense. Never yield to force; never yield to the apparently overwhelming might of the enemy.*[1]

Sir Winston spoke those words in 1941, and they have reverberated down through the decades as a model of an inspirational speech. But the prime minister was using negativity to inspire; he was telling his audience what *not* to do.

Negativity is a difficult form of communication. It has become the campaign method of choice in politics. While it often proves effective—as we saw in Mitt Romney's victorious primary campaign to become the 2012 Republican candidate for president—it leaves a hostile residue and a divided electorate.

In business, negativity fails to provide information. How often have you heard this statement in a presentation?

What we're not is...

Huh? Well then, what *are* you? Tell your audiences what you *are*, not what you are *not*. Moreover, negative statements sound defensive.

One of history's most famous negative statements was President Richard Nixon's infamous defense of himself in the Watergate scandal, "I am not a crook."

Had he framed his statement positively as "I am an honest man," history might have remembered him more forgivingly.

Does this mean that you should never say "never" unless, like Sir Winston, you are exhorting your audience? Mardy Grothe, the author of *Neverisms*, a collection of quotations that begin with the ultimatum "Never," defines Sir Winston's technique as "dehortations," or statements intended to advise against a particular action.[2]

By all means, when you want to inspire, dehort to your heart's content; you will be in good company. In an article about Mr. Grothe's book, Erin McKean, founder of the online dictionary Wordnik.com, extracted some famous dehortations:

- "Never send a boy to do a man's job."
- "Never speak ill of the dead."
- "Never judge a book by its cover."
- "Never count your chickens before they're hatched."
- "Never make the same mistake twice."[3]

My personal favorite dehortation was coined by Leroy "Satchel" Paige who, after a lengthy career in the Negro Leagues, became the oldest rookie—at 42—in Major League Baseball after Jackie Robinson had broken the color barrier. When asked about how he was able to stay youthful and competitive, Mr. Paige said, "Don't look back, something may be gaining on you."[4]

However, in business, negativity for negativity's sake brings problems to the forefront and can lead a presentation into a black hole—the "Houston, we've got a problem!" problem.

Instead, focus on the upbeat, the potential, the road ahead, the actions you are taking, the vision that propels you.

This is not to say that you should sweep problems under the rug or ignore the elephant in the room; you must always be accountable and tell your full story. (You'll see further advice about accountability in Chapter 65, "Breaking into Jail.") Just be sure that, if you bring up the negative, you balance it with the positive.

As the old World War II song advised, "Accentuate the positive."

15

From Bogart to Gingrich

Who Did It?

One of the best moments in the classic 1942 film *Casablanca* is a scene between Rick, the owner of a nightclub in Casablanca, played by Humphrey Bogart, and Captain Renault, the commander of the French forces in the city, played by Claude Raines. At the time of the film's setting, during the early days of World War II, allegiances and political positions were, at best, guarded. Each of the characters in the film has a skeleton in the closet, and each of them is reluctant to reveal any information—as this exchange between Rick and Captain Renault demonstrates:

> Captain Renault: *What in heaven's name brought you to Casablanca?*
>
> Rick: *My health. I came to Casablanca for the waters.*
>
> Captain Renault: *The waters? What waters? We're in the desert.*
>
> Rick: *I was misinformed.*[1]

The exchange also demonstrates the passive voice, the form of speech in which the doer of the action is not stated. Who misinformed Rick? And how could he possibly have come to the desert seeking waters? This verbal dodge, so cleverly crafted by screenwriters Julius J. Epstein, Philip G. Epstein, and Howard Koch, serves Mr. Bogart's character and any deep dark motives or secrets he wants to conceal, but it doesn't serve presenters because this form of speech sounds evasive.

In certain spheres, the passive voice is actually the preferred form of expression:

- **Science.** To avoid ego and maintain objectivity. The phrase, "Further analysis showed..." does not identify who did the analysis.
- **Social.** To avoid direct conflict. "An offense was taken..." Who took the offense?
- **Politics.** To avoid responsibility: "Mistakes were made." Who made the mistakes?

In 1987, President Ronald Reagan, who had avoided commenting on a sensitive political subject, the involvement of his administration in the Iran–Contra scandal, finally agreed to address the subject in a press conference. The headlines in the next day's newspapers carried his key statement: "Mistakes were made."

But the president didn't say *who* made the mistakes.

Ten years later, President Bill Clinton, who had avoided commenting on a sensitive subject, the involvement of his administration in improper fundraising activities, finally agreed to address the subject in a press conference. The headlines in the next day's newspapers carried his key statement: "Mistakes were made."

But he didn't say *who* made them.

Ten years later, Attorney General Alberto R. Gonzales, in a press conference on a sensitive political subject: the dismissals of eight federal prosecutors, acknowledged that "mistakes were made."

But he didn't say *who* made them. History repeats itself.

In 2011, former House Speaker Newt Gingrich, in anticipation of announcing his campaign to become the 2012 Republican presidential candidate, sat down for an interview with the Christian Broadcasting Network. In response to a question about his earlier extramarital affairs, Mr. Gingrich replied, "There's no question at times of my life, partially driven by how passionately I felt about this country, that I worked far too hard and *things happened in my life* (Italics mine) that were not appropriate."[2]

In his use of the passive voice, Mr. Gingrich shifted the responsibility for his inappropriate things—his extramarital affairs—away from himself and to his work.

The public has learned to endure such equivocations from politicians. Mr. Reagan, Mr. Clinton, Mr. Gonzales, Mr. Gingrich, and generations of politicians before and after them use the passive voice to protect their associates, their constituencies, and themselves. In business, where *accountability* is paramount, sentences formed in the passive voice are *not* acceptable; they remove the doer of the action, and with it, remove the presenter from any responsibility or culpability for the action, whether bad or good.

In a *New York Times* article about the passive voice in politics, William Schneider, a political advisor, noted, "...that Washington had contributed a new tense to the language. 'This usage,' he said, 'should be referred to as the past exonerative.'"[3]

This is not meant to be a lesson in syntax, but a lesson in psychology. The difference between the passive voice and the active voice is subtle in grammar, but profound in impact in speech. Avoid the former; use the latter. Put the doer into your sentences.

Use the active voice and become a man or woman of action.

16

Rupert Murdoch's 90% Apology

Who Did It?

In 2011, when a mobile phone hacking scandal rocked the foundation of his media empire, News Corp. Chairman and CEO Rupert Murdoch issued an apology in full-page newspaper advertisements the day after the story broke. The ads, headed "We are sorry," went on to read:

> *We are sorry for the serious wrongdoing that occurred. We are deeply sorry for the hurt suffered by the individuals affected. We regret not acting faster to sort things out. I realise that simply apologising is not enough.*[1]

At first glance, the apology appears to take full responsibility—but not quite. Mr. Murdoch composed that first sentence, "We are sorry for the serious wrongdoing that occurred," in the passive voice, in which the doer of the action is not stated. This is a form of speech common in British culture (as common as the British spelling style of "realise" and "apologising") to convey their characteristic reserve; but it is also form of speech used by public figures—particularly politicians—to duck responsibility, as you saw in the previous chapter.

The sentence does not say who did the wrongdoing. If Mr. Murdoch had used the active voice and written, "We are sorry for the serious wrongdoing *we* committed" instead of "that occured," he would have taken 100% responsibility.

As an Australian, Mr. Murdoch is steeped in British culture. He is also a lifelong journalist who is fully schooled in grammar. But he is also a man who was faced with an embarrassing public situation aptly described by the *New York Times* headline "Tentacles of Phone-Hacking Scandal Grow Tighter,"[2] and he had to tread very carefully.

Which raises two questions: Which aspect of Mr. Murdoch's character did his newspaper apology reflect, and what was his true intent?

Flash forward to 2012 when, after a nearly year-long investigation by the United Kingdom Parliament's Culture, Media and Sport Select Committee, the group issued a report that found that Rupert Murdoch is "not a fit person to exercise the stewardship of a major international company."[3]

Humbled, Mr. Murdoch sent an e-mail to his News Corp. employees saying, "We certainly should have acted more quickly and aggressively to uncover wrongdoing. We deeply regret what took place and have taken our share of responsibility for not rectifying the situation sooner."[4]

He finally took full responsibility with the active voice and made a 100% apology.

17

Winning and Losing the World Cup

He's Just Not That into FIFA

Former President Bill Clinton has the undeniable gift of gab. He won two election campaigns and talked himself out of the Monica Lewinsky mess with only an impeachment and no conviction. In the time since he left office, Mr. Clinton has earned millions of dollars in speaking fees.[1]

He is also widely sought after as a spokesman for not-for-profit causes ranging from raising funds for the victims of the Haitian earthquake and the Indian Ocean tsunami to stumping for Democratic candidates in elections. Even President Obama, a formidable speaker himself, invited Mr. Clinton to deliver a supportive keynote at the 2012 Democratic National Convention.[2]

So it was no surprise that Mr. Clinton was asked to pitch the United States' bid to host the World Cup in 2022. But given his performance of the speech to FIFA, the governing body of the World Cup, it was no surprise that FIFA denied his bid and awarded the plum to Qatar. NBC Sports reported, "instead of the magnetism that defined his eight years in the White House, Clinton sounded dry, reading off his script."[3]

Worse still was Mr. Clinton's script itself.

His speech (available on YouTube) ran 13 and a half minutes, but Mr. Clinton spent most of it focused on his own point of view, including his daughter's experience of playing soccer as a child, his own charitable foundation, the Clinton Global Initiative, and the diversity of sports in America. He finally got around to focusing on FIFA around the 10-minute mark, saying, "FIFA is the main reason that soccer has become a unifying force in the world." Then, referring to FIFA's social responsibility commitment, he added, "If you come to the United States, we'll make sure you fulfill that commitment."[4] Too little, too late.

Steve Kelley in the *Seattle Times* summed up the imbalance, "former President Bill Clinton talked a little too much about, well, Bill Clinton."[5]

In sharp contrast, the winning bid for Qatar was shepherded by a British professional sports consultant named Mike Lee who had helped London win the 2012 Olympics and Rio win the 2016 Summer Games. Mr. Lee's secret weapon is one available to everyone who presents: Put more emphasis on the audience than yourself.

According to a profile of Mr. Lee in London's *Guardian*, the London victory came about because, "He recognised what buttons needed pushing in the often remote International Olympic Committee." Mr. Lee also recognized which FIFA buttons to push, as he told the *Guardian*, "You have to focus on the voters and what the narrative means to them. This is their crown jewel and you have to show how you will take it forward for them."[6]

Winning presenters focus on their audiences. If you're not that into your audience, they just won't be that into you.

18

John Doerr's "Chalk" Talks

Three Best Practices from a Top Venture Capitalist

John Doerr, a partner at Silicon Valley venture capital firm Kleiner Perkins Caufield & Byers (KPCB), is in great demand as a speaker. His repute is attributed to his diverse and successful involvements in for-profit companies (Google and Amazon), not-for-profit organizations (NewSchools Venture Fund), and public policy (The President's Council on Jobs and Competitiveness). Mr. Doerr is often invited to share his experiences, insights, and best practices, and he does so in an unorthodox way: Rather than stand and deliver from a canned deck of PowerPoint slides, he asks his audiences what they want to hear and then fulfills their requests.

Whenever Mr. Doerr steps up to the front of a room, be it a conference center stage or a university auditorium, he polls his audience for the subjects they'd like him to address. He annotates their requests on a large whiteboard—his version of the classic academic "chalk" talk—then proceeds to discourse on each subject. Of course, leaving nothing to chance, Mr. Doerr researches the audience, the organization, and the event, enabling him to anticipate the key themes he might be asked about. To support his discussion, he brings along a few

PowerPoint slides to illustrate the themes, and he accesses the slides as he makes his way through his whiteboard list.

In doing so, he provides a role model for three important presentation best practices:

1. Elevate the audience's primacy. One of the most common faults that presenters make is to sell features rather than benefits; they focus on their message without regard for the audience (witness the one-size-fits-all "Corporate Pitch"). The usual result is a failure to achieve the goal of the presentation. Mr. Doerr's approach rights the balance.
2. Relegate the slides to their proper secondary role. Undoubtedly, another common fault is to multitask the role of slides; presenters use them not only as illustrative graphics, but also as speaker notes, send-aheads, and leave-behinds. This approach produces images of encyclopedic detail that serve none of the functions. Here, too, Mr. Doerr's approach rights the balance.

   Moreover, in accessing his slides randomly, he employs a useful, but little-known, Microsoft PowerPoint technique:
3. The "Go To" Command. When PowerPoint is in Slide Show mode, Mr. Doerr—or you—can go directly to any slide in the deck by entering the slide number (prompted from a printed outline of all the slides) and pressing the "Enter" key. These simple strokes will jump the slide show directly to the desired slide.

The last technique has three benefits:

- The presenter appears in complete command and control, sending the subliminal message that the presenter is managing the event effectively. Management is the primary decision factor for investors.
- Instant gratification for the audience, nice to have for any human being, vital for every audience.
- The "cool" factor.

In his primary role as a venture capitalist, Mr. Doerr sees many presentations from many companies that pitch him to invest many millions of dollars. Surely, he measures what he sees and hears through the filter of his own best practices.

How would you measure up?

19

Vinod Khosla's Cardinal Rule

"Message Sent Is Not the Same as Message Received"

By any measure, venture capitalist Vinod Khosla is one of the most influential people in business today. In his long and distinguished career, Mr. Khosla has contributed to the growth of hundreds of companies, first at the renowned venture firm of Kleiner Perkins Caufield and Byers, and then, since 2004, at his own firm, Khosla Ventures. Among his notable successes are Sun Microsystems, Nexgen/AMD, Excite, and Juniper Networks. On their way to maturity, each of the many companies Mr. Khosla touched came under the scrutiny of his expert eye, assessing their business plans, balance sheets, strategic relationships, marketing materials, and especially their presentations.

To all of them, he applied his cardinal rule of communication: "Message sent is not the same as message received," an eloquent statement of the obligation of all presenters to ensure that their target audience has received the intended message.

Fulfilling that obligation requires a full court press that can be described, with all due respect to Stephen Covey, as "The 7 Habits of Highly Effective Presenters":

1. **Analyze your audience in advance.** Just as salespeople qualify their customers, presenters must qualify their audiences. In your preparation, gather as much information as you can about who they are, what they know, and what they want to know; identify their concerns, fears, and hot buttons.
2. **Develop focused content.** Armed with your thorough analysis, create content that addresses your audience's interests. An essential part of this process is to eliminate irrelevant information—easier said than done because most presenters operate under the assumption that, for their audience to understand anything, they must tell them everything. Wrong! Tell them only what they need to know.
3. **Offer multiple benefits.** Infuse your pitch with benefits. Find multiple points in your presentation where you can insert a sentence that begins, "The reason this is important to you..." and then concludes with a benefit to your audience.
4. **Customize, customize, customize.** In today's high pressure, high stakes business world, presenters—who have become time-constrained road warriors—try to save time by making a one-size-fits-all pitch. Wrong again! Use the information you collect about your audience in your preparation and insert references throughout your presentation. Keep it fresh and specific.
5. **Track your progress as you present.** The importance of eye contact in presentations is a given, but most presenters merely scan their audiences and see nothing. Read your audience's reaction to your story. Look for their head nods. See if your message is being received.
6. **Adjust your content.** If instead of head nods, you get frowns or puzzled looks, pause in your narrative and add a brief explanation, or ask your audience if they have questions.
7. **Respond to all questions in full.** Whether you get questions during or after your presentation, you—unlike politicians—must respond. This is not to say that you must reveal strategic or confidential information, but you must address the issue in every question and give a reason when you cannot. Never evade.

Although Mr. Khosla applies his cardinal rule to the presentations of his existing portfolio companies—and those who aspire to become one of his portfolio companies—he represents every member of every audience of every presentation you will ever give. If you aspire to succeed, make sure that every message you send is received—loud and clear—by every audience.

20

The Outline Trap

Britannica and Brainstorming

One of the early lessons we all learn in school is how to make an outline;[1] how to create that waterfall of Roman numerals, capital letters, Arabic numerals, and lowercase letters that cascade down to the bottom of the page, if not dozens of pages of interminable term papers. Thus we are programmed to arrange our ideas in a hierarchical order—in sharp contrast to what our brains do naturally: generate ideas in random order.

To demonstrate: As I sit writing this chapter, I glance at a ballpoint pen on my desk. The logo on the pen reminds me that I got it at as souvenir at a conference. I remember that I met a man at the conference who told me about a book on presidential politics. This reminds me that I had been planning a blog on the same subject, and so I open a file with the notes on that subject and... you see where this is going.

I'm sure that if you were to track your own thought patterns, you would discover the same winding, random path. That's the way every human mind works: unstructured.

And yet, when businesspeople sit down to develop a presentation, they immediately start to apply structure, in either a hierarchical outline form or by organizing a set of existing PowerPoint slides to create a new "deck"—each approach forces structure onto unstructured ideas.

In Chapter 6, "How Woody Allen Creates," you read that he and other artists let their random ideas flow unimpeded, capture them in notes as they occur, and then lay out the notes in a panoramic view. Mr. Allen tosses scraps of paper onto his bed, other film directors use storyboards, architects make papier-maché models, military officers use wall size maps, and businesses encourage employees to doodle their creative ideas on whiteboards during product development or strategy sessions. The *Wall Street Journal* reported that sales of IdeaPaint, a paint product that turns a wall surface into a whiteboard, have doubled since 2008.[2]

For your presentation development, you can do your brainstorming on a whiteboard, a computer screen, or Post-it Notes as you generate your ideas, but what is as important as the free flow is that you see the ideas you generate in a panoramic or landscape view.

The simple reason for this aspect of the creative process is that our eyes are set side-by-side in our heads, making the landscape view more pleasing and open than the portrait view. If you start with an outline, the constricted view imposes a ranking sequence too early in the process. A panoramic view allows you to see the conceptual relationships among your ideas.

Even the venerable Encyclopedia Britannica has come to understand the importance of visual thematic relationships. The publisher of alphabetized—sequential rather than conceptual—reference works for almost 244 years, discontinued its print version in 2012 and went digital. As part of the transition, they included a link map feature that looks like a brainstorming session you might do on a whiteboard.[3]

Walter S. Mossberg, the author of the *Wall Street Journal's* "Personal Technology" column, reviewed the feature and wrote that the publisher, which "has always been expensive, and a bit stodgy...has produced a slick app.... Perhaps the coolest feature is the link map, triggered from an icon at the top of each article page. This generates a spider web of icons representing other articles related to the one you were reading."[4]

The 35,000 foot view shows patterns that lead to clear stories; an outline traps ideas.

Take the high road.

21

Having a 'versation

"I" Versus "You"

There's an old joke about the opera diva who receives an adoring fan in her dressing room after a performance. The diva goes on and on about how magnificently she sang her arias, about her dramatic acting, her expressive gestures, and her fabulous costumes. After about half an hour, the diva says to the fan, "But enough about me, what did *you* think of my performance?"

Cartoonist Joe Dator did a variation on the diva joke for *New Yorker* magazine. In the sketch, a man is speaking to a woman seated across a table. The caption reads, "Enough about me, but nothing about you just yet."[1]

This is no laughing matter in most other walks of life, for self-centeredness creates a high obstacle to all communication. In presentations, self-centeredness is manifested by a lack of relevance to the audience, and in sales by the lack of benefits for the customer. But to fully understand the negative impact of one-way communications, let's focus on the more universal view offered by interpersonal exchanges. The benchmark for conversation was set by Sherry Turkle, a psychologist and professor at M.I.T., in an article in the *New York Times:*

> *In conversation we tend to one another. (The word itself is kinetic; it's derived from words that mean to move, together.) We can attend to tone and nuance. In conversation, we are called upon to see things from another's point of view. Face-to-face conversation unfolds slowly. It teaches patience.*[2]

Sadly, the opposite has become the norm today in what is known as "Having a 'versation."

We've all been trapped by party bores who emulate the opera diva by delivering monologues all about themselves. One of the early indications that the one-way street is heading for a dead end is the ratio of declarative statements to questions. Bores speak with no question marks on their verbal keyboard.

Another indicator is the ratio of how frequently bores say "I" to how *in*frequently they say "you." That simple metric serves as an early warning for you to excuse yourself and head for the bar to refresh your drink. But the role of pronouns in communication extends beyond chit chat into interpersonal relationships.

Another psychologist, James W. Pennebaker, Professor and Chair of the Department of Psychology at the University of Texas at Austin, studies the connections between the frequency of words and feelings. In his book, *The Secret Life of Pronouns*, he writes:

> *Pronouns (such as I, you, we, and they)...broadcast the kind of people we are.... By looking more carefully at the ways people convey their thoughts in language, we can begin to get a sense of their personalities, emotions, and connections with others.*[3]

Professor Pennebaker conducted a variety of research projects ranging from Craigslist ads to Twitter messages to prove his point. One of the most revealing was a study on speed-dating, which, according to a report in the *New York Times*, "found that couples who used similar levels of personal pronouns, prepositions and even articles were three times as likely to want to date each other compared with those whose language styles didn't match."[4]

This chapter is not meant to help you improve your results at speed-dating, but to urge you to match closely with your listeners, to focus on the "co-" in *co*mmunications, to have *con*versations, not 'versations.

When you present, be mindful of your audience by offering them benefits; when you converse, be mindful of the other person by balancing your "I" to "you" ratio. When in doubt, err on the side of the latter.*

* Professor Pennebaker offers an opportunity to assess your compatibility with a friend by tracking your word usage in this online exercise: www.secretlifeofpronouns.com/exercise/synch.

22

"It's all about you!"

"...But they're just not that into you."

In the previous chapter, you read about how self-centeredness is an obstacle to all communication, extending all the way from social conversations to our focus, presentations. To remove that barrier, to put the "co-" in "communication," the effective communicator adds interaction to interpersonal exchanges, but more important, adds benefits for the listener—whether that listener is the other person in a conversation or the audience for a presentation.

But that leaves open the question of why any person in his or her right mind would allow a failure to communicate to occur in the first place. There are two answers: one scientific and the other a pervasive misconception that has taken on the status of a legacy in the world of presentations.

Harvard neuroscientist Diana Tamir and her Harvard colleague Jason Mitchell conducted a series of experiments to explore why people like to talk about themselves. The results, published in the Proceedings of the National Academy of Sciences, said:

> *Here, we test recent theories that individuals place high subjective value on opportunities to communicate their thoughts and feelings to others and that doing so engages neural and cognitive mechanisms associated with reward. Five studies provided support for this hypothesis. Self-disclosure was strongly associated with increased activation in brain regions that form the mesolimbic dopamine system, including the nucleus accumbens and ventral tegmental area.*[1]

The mesolimbic dopamine system just happens to be the same part of the brain in which pleasurable sensations occur. Figure 22.1 shows activation of the brain during self-disclosure. Meaning that, as the *Wall Street Journal* story about the study summarized it, "Talking about ourselves—whether in a personal conversation or through social media sites like Facebook and Twitter—triggers the same sensation of pleasure in the brain as food or money."[2]

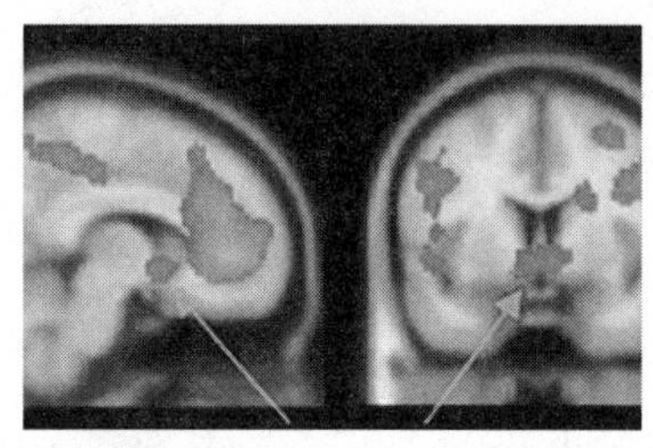

Figure 22.1 Neural response during self-disclosure

This places a very high barrier to being able to interrupt the party bores who monopolize conversations. The only solution I can offer is to repeat the one from the previous chapter: excuse yourself and head for the bar to refresh your drink.

Presentations are another matter. As a coach, I have spent the greater part of my career urging presenters to include benefits in their pitches. But the need to remind them still persists. Presenters continue to sell features and/or blow their own horn. The reason they do—and this is only conjecture—goes all the way back to the dawn of the presentation universe when some sage decided that presentations should begin with a "snapshot" that introduces the presenter's company. This usually results in an initial slide that, depending on who creates it, is a hodgepodge of disparate facts that including but not limited to:

- year of founding
- number of employees
- value proposition
- financial results
- markets served
- key customers
- office location
- square footage

This step gets the presentation off on the wrong foot for a number of reasons: The slide attempts to tell the whole story, the story is not apparent at a glance, the focus shifts attention away from the presenter, the presenter is forced to read the slide... the list goes on. But worst of all, it's all about you and not about the audience and, to paraphrase the title of the 2004 bestselling book about dating, they're just not that into you.[3]

Make the front end of your presentation about your audience. Focus on their issues and concerns and tell them what your company can do for them. Pivot from your point of view to theirs. This pivot is best illustrated by the story of Theodore Leavitt, a professor at Harvard Business School who told his students not to try to sell customers a quarter-inch drill, but a way to make a quarter-inch hole.[4] Tie what you do to your audience's needs.

Consider the snapshot slide as boilerplate that is best left to the handout materials. If you still feel the need to include information about your company within the presentation, shift it to later in the deck, *after* you have shown them how well you understand them.

It's all about them.

23

When *Not* to Tell 'em

"Get on with it!"

Senator Barry Goldwater, the Republican candidate for president in 1964 infamously said, "Extremism in the defense of liberty is no vice. And moderation in the pursuit of justice is no virtue."[1] This view of political policy became a major factor in Mr. Goldwater's landslide loss to Lyndon Johnson, but it also serves as a warning lesson for presenters. Extremism in any pursuit can overshoot the mark and result in the opposite intent of the pursuit.

One of the most frequently repeated pieces of advice for presenters is to "Tell 'em what you're gonna tell 'em, tell 'em, and then tell 'em what you've told 'em." I offer the same advice in my own coaching practice and writing. The intent is to impose and maintain a clear narrative flow in presentations and speeches, and the reason it is repeated so often is that most presenters and speakers, who regularly crank out long, rambling, pointless patchwork pitches, desperately need reminding. The Triple "Tell 'em" is one solution. However, too much of a good thing can be a bad thing; a sword can cut two ways.

Bruce Eric Kaplan, a cartoonist who appears regularly in the *New Yorker* magazine as BEK, skewered the excessive Triple "Tell 'em" in one of his panels that showed a presenter in front of an audience, saying, "First, I want to give you an overview of what I will tell you over and over again during the entire presentation."[2]

We're also painfully familiar with presenters who impose a narrative laundry list on their bullets by saying "First, I'd like to talk about..." then move on to the second bullet saying, "Next, I'd like to talk about..." and then proceed through every bullet the same way until the very end, when they say—wait for it— "Last but not least...."

Some presenters push their extreme handholding even further, by utilizing their slides to do the tracking. As in Figure 23.1, they insert copies of an agenda slide between the sections of their presentation, progressively shifting the highlighted bullet to "Tell 'em what they're gonna tell 'em" in the upcoming section.

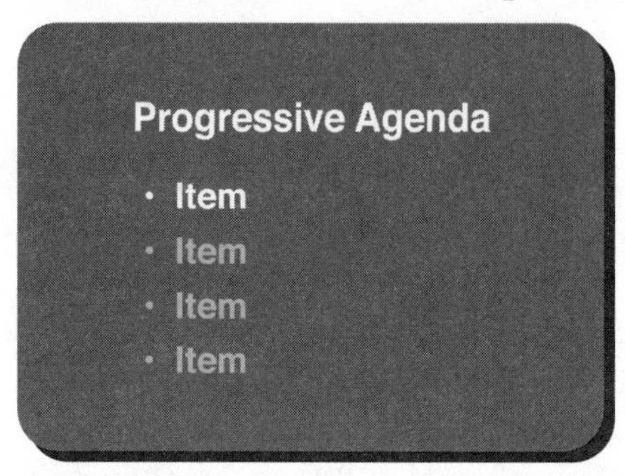

Figure 23.1 Progressively highlighted agenda

This technique can be useful in long tutorial presentations, but if there are only one or two slides between the variations of the agenda in short presentations—and short presentations are obligatory in this Twitter-paced day and age—the audience, feeling patronized, will react with a big *Duh*!

Presenters are not the only perpetrators of such deliberate continuity devices. Geoff Dyer, who writes the "Reading Life" column in the *New York Times* Book Review section, considers excessive tracking a "basically plodding method." In one of his columns, he criticized art historian Michael Fried, whose book *Why Photography Matters as Art as Never Before* takes "the style of perpetual announcement of what is about to happen to extremes." Mr. Dyer said it is "like watching a rolling news program: *Coming up on CNN... A look ahead to what's coming up on CNN.*" Concluding his critique, Mr. Dyer wrote, "I kept wondering why an editor had not scribbled 'get on with it!' in huge red letters on every page of the manuscript."[3]

To keep your audience from thinking "Get on with it!" apply the Less is More rule, not just in your slide design but in your content. As you've read repeatedly throughout this book:

- Edit the amount of material you present.
- Be brief *and* concise.

Then, with a shorter and more succinct story, look at your presentation from a 35,000 foot view—as a storyboard—in the Microsoft PowerPoint Slide Sorter view, or with the Power Presentations Storyboard form in Figure 23.2.

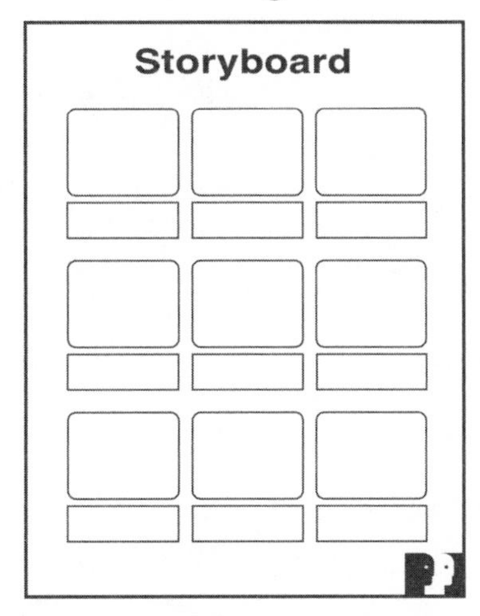

Figure 23.2 Power Presentations storyboard form

The storyboard is downloadable from our website, www.powerltd.com, by clicking at the bottom of the home page.

Just as television and film directors use storyboarding to see the full scope of their stories, look at your slide show in this panoramic view to see your flow. Then rehearse your presentation *aloud,* moving from frame to frame. Do this several times. Along the way, you'll find that you might want to add, delete, or shuffle slides. As you proceed with your iterations, you will develop verbal connective links for your narrative.

Ultimately, you will have a presentation in which The Triple "Tell 'em" is transparently implied. You will have a story that is easy for you to deliver and, more important, easy for your audience to follow—without a laundry list, without CNN-style teasers, and best of all, without those patronizing agenda slides.

Get on with it!

24

Bookends

Establish Your First and Last Sentences

Novelist John Irving, the author of the bestselling *The World According to Garp*, *The Cider House Rules*, and *The Hotel New Hampshire*, offers two valuable pieces of advice to writers that presenters would do well to follow. On the occasion of the publication of his thirteenth novel, *In One Person*, he wrote, "I begin with endings, with last sentences—usually more than one sentence, often a last paragraph (or two). I compose an ending and write toward it, as if the ending were a piece of music."[1]

By beginning the development of your presentation with the last sentence—the one in which you make your call to action, just before you open the floor to questions—you give your entire narrative a focus and forward thrust.

Mr. Irving also offered advice that addresses the front end of your presentation. The late William Safire, the revered literary critic and author of the long-running "On Language" column in the *New York Times*, wrote, "There is a tough assignment about openings from John Irving: 'Whenever possible tell the whole story of the novel in the first sentence.'"[2]

That's a rather large challenge for businesspeople who have not published 13 novels, but it sets a higher bar for those presenters (and you know who you are) who start their stories at the Dawn of Civilization and work their way forward to the present, step-by-excruciating step. Take six or seven sentences, if you must, but make sure that you front load your presentation with a succinct progression that achieves the following steps:

- Captures their attention
- Describes your value proposition
- States your point or call to action

Readers of *Presenting to Win* will recognize these as the key steps of the Opening Gambit, a sequence of sentences designed to hook your audience at the very beginning of your presentation. You never get a second chance to make a first impression.

Please note that both of Mr. Irving's practices relate to getting to the point, and these mirror images form bookends for your presentation. Good advice, from a man who knows a thing or two about books.

Mr. Irving's words about strong openings were echoed by fellow-novelist Jhumpa Lahiri, the winner of the 2000 Pulitzer Prize for Fiction and the author of the bestselling *The Namesake*. She wrote, "The first sentence of a book is a handshake, perhaps an embrace."[3]

Embrace *your* audience.

25

The Sound of *Ka-Ching*!

Scale the "You"

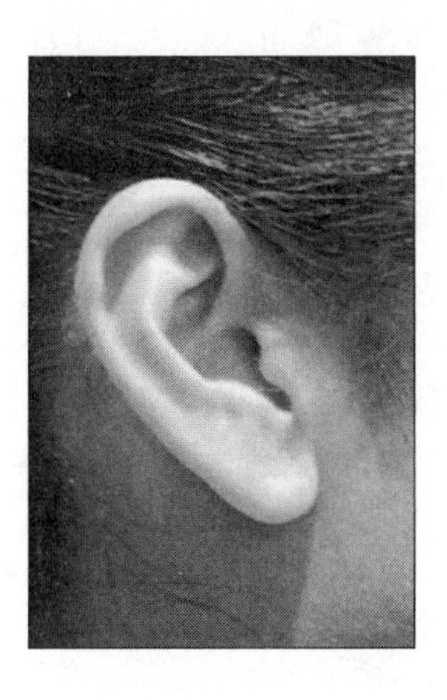

If you've ever tried to have an important conversation on your mobile telephone from a busy airport terminal, a crowded restaurant, or a noisy street, you'll appreciate the solution developed by Audience, Inc., a Silicon Valley company. Using sophisticated software algorithms and integrated circuits that mimic the dynamics of the human ear, Audience's technology separates the background noise from the human voice and transmits conversations with crystal clarity. Investors appreciated Audience's technology—and the 316% Compound Annual Growth Rate of their revenues—when they went public in May 2012. The stock was priced above the anticipated range,[1] and then rose 12% higher on the first day of trading.[2]

Investors related to Audience as end users of mobile devices, but they also related in their primary roles as investors because Peter Santos, the company's CEO, made the connection for them in his IPO road show. After describing Audience's revolutionary solution—with a simple but dramatic demonstration of how the technology enhances a mobile conversation on a noisy street—Mr. Santos punctuated his presentation by scaling the "you."

If you or billions of other mobile phone users want to have not just narrowband calls but wideband calls, to be able to have video in addition to voice calls, want to be able to enter addresses or search strings using your voice instead of typing it on a keypad, and if you want to do all these things in a noisy mobile environment, you have a much higher bar in terms of acoustic signal processing.... Audience has solved those problems.

Because Mr. Santos scaled from the single "you" of a mobile phone user to the implied multiples of "you," represented by the "billions of other mobile phone users," he demonstrated the large market potential for Audience's technology. But he didn't stop there, he went on to widen the opportunity for that same technology to address several other applications and several other markets; all of which is music—as clear as an enhanced mobile phone call—to investors' ears, music accompanied by the sound of a ringing cash register: *Ka-Ching*!

Scaling the "you" was a lot easier for Mr. Santos than it was for the CEO of a medical device company whose product treats diabetes victims. In her IPO road show, the CEO could not address the investor audience as individuals for two reasons: less than 10% of the population has diabetes[3] and, far more important, she might have made her audience feel uncomfortable by associating them with a debilitating disease. In her road show she went right to the big market by describing the 25.8 million children and adults in *just* the United States who could be helped by her company's product. *Ka-Ching!*

The CEO of another medical device company, whose product is a minimally invasive surgical tool, had it a bit easier. I coached his IPO road show and, during our sessions, role-played a potential investor. In the final rehearsal of his presentation, the CEO held up his minimally invasive surgical device, looked at me and said "With this device, you can make a better incision." I shook my head and said, "I don't make incisions." He thought for a second, held up the device again and said, "So you can see, if tens of thousands of surgeons want to make better incisions, they're going to have to buy this device from us." *Ka-Ching!*

Do the calculations for your audiences and add value.

Scale the "you," and listen for the rewarding sound of *Ka-Ching!*

26

David Letterman's Top Ten

Pick a Number

In his nearly two decades as the host of CBS's *Late Show*, David Letterman has made his nightly reading of his "Top Ten"[1] list a social ritual of American culture. Mr. Letterman uses his list for comic effect, but you can use the same approach to create a structure for your presentations.

Authors Stephen R. Covey and Deepak Chopra used the numbering technique as the structure of their respective bestsellers, *The 7 Habits of Highly Effective People* and *The Seven Spiritual Laws of Success*. The popular Politico website[2] regularly calls out their Top Five or Six or Seven takeaways from the political events they cover; how-to newspaper and magazine articles add sidebar boxes that summarize their main tips with a total number; and the help desk web page of product and service companies summarize their customer FAQs with a total number.

In the fast and furious business world where presentations are often hastily cobbled together with a disparate collection of begged, borrowed, or stolen slides and delivered by a presenter who is the only one in the room who can understand what on Earth one slide has to do with another, the numbering technique can be emergency CPR. Simply organizing the different elements into a clear order makes it easy for both the presenter and the audience to follow.

Eric Benhamou, the former Chairman of 3Com Corporation (acquired by HP in 2010), did so under rather trying circumstances.

Mr. Benhamou was invited to deliver a keynote speech at a dinner given by the California-Israel Chamber of Commerce, an organization as diverse as the more than 7,000 miles that separate those two centers of business. The event, which was held on a midweek night at Silicon Valley's large San Jose Fairmount Hotel, began with a cocktail hour that ran for far more than an hour. When the ballroom doors finally opened, the several hundred guests rushed in to find seats at tables they had to share with strangers. After the usual rubber chicken dinner, the Masters of Ceremonies presented awards to individuals who were only familiar to Californians, and some who were only familiar to Israelis. Each of the recipients then proceeded to give an acceptance speech that made Academy Award acceptance speeches seem abrupt by comparison. When Mr. Benhamou's turn came, it was nearly nine o'clock.

How would you like to have to deliver a speech in those circumstances?

But Mr. Benhamou rose to the occasion. At the very outset, he announced that he would be sharing the top ten experiences of a recent trip he made to Israel. Each of the experiences was unrelated to the next—one historic, one cultural, one economic, one technological, one a human interest story—but he counted down as he went from anecdote to anecdote. Mr. Benhamou is a gifted presenter, and he held his audience's attention throughout. When he got to his ninth anecdote, the audience began reaching for their valet parking stubs and, as soon as he finished the tenth, they bolted for the doors.

Structure your presentation by organizing your diverse themes into a group of ten or seven or three—many people subscribe to what is commonly known as the "rule of threes." Give your audiences order rather than chaos, or they will start reaching for their valet parking stubs a lot sooner than Mr. Benhamou's did.

27

Illusion of the First Time

Road (Show) Warriors

Suit the action to the word, the word to the action.

—*Hamlet*, Act III, Scene 2,
William Shakespeare

As a presentation coach, I draw an indelible line between presenting and acting; this despite the fact that my education includes a Master's degree in Speech and Drama from Stanford University. I do this because the businesspeople I coach—already stressed about the mission-criticality of their presentation—blanch at the thought of having to perform.

Nonetheless, I have carried forward one lesson from my studies in drama. It comes from William Gillette (1853–1937), an American actor whose claim to fame was his portrayal of Sherlock Holmes, a role he played more than 1,300 times. No presenter is ever going to tell the same story 1,300 times, but the road warriors who do multiple iterations of their pitch—to sell products, seek partners, or solicit financing—face the challenge of keeping each of those iterations fresh.

Mr. Gillette met that challenge by creating what he called, "The Illusion of the First Time in Acting," a subject he described in his book of the same name:

> *There yet remains the Spirit of the Presentation as a whole. Each successive audience before which it is given must feel—not think or reason about, but feel—that it is witnessing, not one of a thousand weary repetitions, but a Life Episode that is being lived just across the magic barrier of the footlights.*[1]

One of the best examples of a pitch that presenters must deliver multiple times is the IPO road show. Conventionally, when a company goes public, the senior management team goes on the road for about two weeks, during which they visit potential investors in about a dozen cities across the country (and often across the oceans as well), telling the same story several times a day, or about 30 or 40 times each week for a total of 80 or more iterations. (Although, as you'll read in Chapter 74, "Ten Best Practices for the IPO Road Show," those investor meetings have changed in one important aspect.) With that kind of schedule, presenters occasionally slip into autopilot and experience the "If this is Tuesday, it must be Belgium" phenomenon—leaving their audiences uninvolved, unmoved, and unconvinced.

Investment bankers, who have worked on as many, if not more, road shows than presentation coaches and have seen management team after management team go into autopilot, have developed a device to avoid that phenomenon. Before each presentation, they pick an unusual word and ask the presenting team to find a way to work it into the presentation. This may keep the presenters alert, but means nothing to the investors.

A more effective technique—available to *any* presenter—is to involve the audience. In an IPO road show, the solution is readily at hand. The investment bankers, who have arranged all the meetings with all the investors, know a great deal about each firm and share that information with the presenters before each session. Unfortunately, in the heat of battle, most presenting teams neglect to use the information.

The lesson for any presenter is to develop knowledge of the audience in advance. In the case of IPO road shows, the spadework is done by the investment bankers, their retail sales force, and their analysts;

other presenters must do it on their own. That means that you must learn as much as you can about each of your audiences before you present. Scour the web for company information, read their latest press releases, see what their industry press, peers, and competitors are saying about them. Visit LinkedIn to learn about the roles and backgrounds of people you will be addressing.

And then use it—or you lose it. Pepper your presentation with the information you have gathered. Think of this technique as a tasteful, appropriate form of name-dropping.

Does this mean that you have to change your recurring presentation each time? *Not at all*. Just add the customized references to the core content of your narrative. You can use this very same technique for a one-time-only presentation, as well as for every presentation you ever give to every audience.

Make your last—especially if it's the eightieth—presentation as fresh as the first. And while you're doing all that preparation, thoroughly rehearse your presentation in advance and make your first iteration as polished as the last.

28

In Praise of Analogies and Examples

Add Value and Dimension

Harvard Business School (HBS) established the gold standard for learning with its case study method. In that curriculum, students learn how to deal with real business situations in participatory exercises. As the school's website explains, "In class—under the questioning and guidance of the professor—students probe underlying issues, compare different alternatives, and finally, suggest courses of action in light of the organization's objectives."[1]

In doing so, HBS is following the precepts coined by the ancient Chinese philosopher, Confucius, who wrote:

I hear and I forget

I see and I remember

I do and I understand

This is not to recommend that presenters become participatory—they already do that every business day of their lives—but that they illustrate their stories with the narrative equivalent of the HBS case study method, by using analogies and examples in their presentations. Professional writers have long used these techniques to illuminate their narratives, but presenters, in their rushed lives, often take a short cut and deliver "Just the facts, Ma'am."

Now there is scientific evidence to support the use of such illustrative material. A *New York Times* article reported that:

> *Brain scans are revealing what happens in our heads when we read a detailed description, an evocative metaphor or an emotional exchange between characters. Stories, this research is showing, stimulate the brain and even change how we act in life.*[2]

The article demonstrated the power of metaphors by citing a study made by a team of researchers from Emory University:

> *[W]hen subjects in their laboratory read a metaphor involving texture, the sensory cortex, responsible for perceiving texture through touch, became active.*

In the early days of the information technology, the superhighway was the metaphor-of-choice; today it is the cloud. The Swiss Army knife symbolizes multitasking; low-hanging fruit is an analogy for a readily accessible opportunity; and Levi's supplying blue jeans to Gold Rush miners represents a first-to-market advantage. Several years ago, one of my pharmaceutical clients described their skin patch drug delivery technology by comparing its action to a truck transporting cargo across a border. Find an analogy that helps illustrate your story.

The *Times* article went on to cite scientific research about the value of case studies or examples. A study made at York University in Canada, found that:

> *[T]here was substantial overlap in the brain networks used to understand stories and the networks used to navigate interactions with other individuals.... Narratives offer a unique opportunity to engage this capacity, as we identify with characters' longings and frustrations, guess at their hidden motives and track their encounters with friends and enemies.*

You can illustrate the value of your business with examples of satisfied customers who bought your products or services, of partners with whom you developed win-win relationships, or of investors who profited from funding your company.

And when you do, put those examples into human terms; use their names. People like to hear about people.

The *Wall Street Journal* has a daily human interest story on its otherwise "just the facts" front page. They call the feature "A-heds," which they describe as "more than a news feature. Ideas rise out of our personalities, our curiosities and our passions."[3]

Both analogies and examples, in addition to activating neural activity in your audiences, add value to your presentations.

In the next chapter, you'll read about two masters of the case study.

29

Ronald Reagan and Barack Obama

Masters of the Game

No book on presentations would be complete without reference to two masters of the game: the 40th and 44th Presidents of the United States, Ronald Reagan and Barack Obama; two men whose politics are poles apart, but who share one common touch point that serves as a lesson for *any* presenter.

Although their speaking styles also differ—Mr. Reagan, the genial former actor from the Midwest, who overwhelmed audiences with his underplaying, and Mr. Obama, the cool former Ivy League attorney, who rouses audiences with his dynamic voice and elegant bearing—both men use their individual styles in the service of their outstanding ability to tell human interest stories.

Mr. Reagan almost singlehandedly invented the anecdotal game. The Great Communicator rarely missed an opportunity to tell a tale about a brave soldier or a dedicated student. Readers of *Presentations in Action* will recall the story of how, in 1983, Mr. Reagan honored the courageous act of a federal employee named Lenny Skutnik by recounting the details of the act during the State of the Union Address—while Mr. Skutnik sat next to Nancy Reagan—establishing a precedent that every president since has followed.

Mr. Obama appreciates Mr. Reagan's talents. In his autobiography, *The Audacity of Hope*, Mr. Obama frequently referenced his predecessor. "I understand his appeal," Mr. Obama wrote, referring to Mr. Reagan's ability to spark Americans to "rediscover the traditional virtues of hard work, patriotism, personal responsibility, optimism and faith. That Reagan's message found such a receptive audience spoke...to his skills as a communicator."[1]

Mr. Obama took his appreciation of Mr. Reagan along with him during his 2010 holiday vacation in Hawaii in the form of a book. At holidays, which are usually slow news periods, media interest sometimes turns to what the president is reading.[2] That year it was a biography called *President Reagan* by Lou Cannon. In it, Mr. Obama read a statement Mr. Reagan made just after he left office:

> *Some of my critics over the years have said that I became president because I was an actor who knew how to give a good speech. I suppose that's not too far wrong. Because an actor knows two important things—to be honest in what he's doing and to be in touch with the audience. That's not bad advice for a politician either. My actor's instincts simply told me to speak the truth as I saw it and felt it.*[3]

Little did Barack Obama know how meaningful that statement would be. Shortly after his return from that vacation, on January 8, 2011, a deranged Jared Lee Loughner shot Representative Gabrielle Giffords and 18 other people during a public citizens' meeting held in a supermarket parking lot in Tucson, Arizona. Four days after that tragic event, Mr. Obama flew to Tucson to address a stunned nation and the families and friends of the victims at a memorial service at the University of Arizona.

After a very brief formal opening of condolences including a passage from Scripture, Mr. Obama began to talk about each of the victims. In simple, but eloquent words, he painted a warm human picture of each person's life—especially that of nine-year-old Christina Taylor Green whose story he extended as a role model for the nation:

> *Imagine—imagine for a moment, here was a young girl who was just becoming aware of our democracy; just beginning to understand the obligations of citizenship; just starting to glimpse the fact that some day she, too, might play a part in shaping her nation's future. She had been elected to her student council. She saw public service as something exciting and hopeful. She was off to*

> *meet her Congresswoman, someone she was sure was good and important and might be a role model. She saw all this through the eyes of a child, undimmed by the cynicism or vitriol that we adults all too often just take for granted.*
>
> *I want to live up to her expectations.*[4]

The passage could have been taken right out of the Ronald Reagan style manual.

Validation came from the *Wall Street Journal's* Peggy Noonan, a former speechwriter for Mr. Reagan and a frequent critic of Mr. Obama. In her postmortem of the Tucson event, she wrote, "About a third of the way through, the speech took on real meaning and momentum, and by the end it was very good, maybe great." She attributed the pivot to "when Mr. Obama started to make things concrete...specific facts about real human beings."[5]

"Specific facts about real human beings," is sound advice for any speaker.

30

Aristotle: The First Salesman

The Original Source

Mortimer Adler, the noted twentieth century professor, philosopher, and chairman of the Board of Editors at Encyclopaedia Britannica, was also a scholar of the classics. In his 1983 book *How to Speak How to Listen,* Mr. Adler described an invitation he received to speak at the Advertising Clubs of California:

> *They asked me in advance for a title. I suggested that it be "Aristotle on Salesmanship," a title I thought would be sufficiently shocking for them. It was. No one had ever before connected the name of Aristotle with salesmanship—or with advertising, which is the adjunct of selling.*[1]

Sadly, no one connects presentations with selling either. Why else, as you read in Chapter 7, "What's Your Point?," do audiences so often mutter to themselves, "What's the point of all this?" or "and your point is?" Such questions are prompted by a pointless story, but more often, by a lack of a call to action.

The answer to those questions may be that, because salesmanship is held in such low esteem in our culture, asking for the order is considered pushy. In a book called *The Art of the Sale*, author Philip Delves Broughton points to two Pulitzer prize-winning plays, Arthur Miller's *Death of a Salesman* in 1949 and David Mamet's *Glengarry Glen Ross* in 1984, as having marked salesmen as archetypically unsavory characters. Mr. Delves Broughton goes on to describe current salesmen as people who are "goaded to perform and reined in when they sell too hard. They are patronized as 'feet on the street' by those who prefer to imagine that business can be conducted by consultants with dueling PowerPoint presentations."[2]

The solution is to go beyond PowerPoint and make your point crystal clear, make an unmistakable call to action. That is the mark of effective salespersons, effective presenters, and effective people in all walks of life. As L. Gordon Crovitz, a former publisher of the *Wall Street Journal*, wrote in his review of Mr. Delves Broughton's book:

> *We all engage in sales of one sort or another. Parents sell the idea of eating vegetables to their children; reporters sell their latest story idea to editors; university presidents sell their institution's neediness to potential donors.*[3]

But asking for order is only part of the solution to effective salesmanship; the other is to provide benefits to customers in sales, and benefits to audiences in presentations. Failure to give benefits in each arena is anathema.

Which brings us right back to Aristotle. 2,300 years ago, the great philosopher proposed that, to be persuasive, a speaker must provide the holy trinity of *Ethos*, or credibility, *Logos*, or evidence, and *Pathos*, or benefits. In the twenty-first century, the first two are givens. Given the curiosity and speed of our media, no speaker or businessperson can get away with un*ethical* behavior or shoddy evidence—for very long. Consider the current parade of exposed politicians and executives whose misdeeds make the sleazy salesmen in *Glengarry Glen Ross* look like kindergarten pranksters.

But pathos is still sadly missing. So if you take away only one lesson from this section on the art of telling your story, it is this: Load every presentation you ever give with benefits for every audience. And do it with this simple rule of thumb: Pause every couple of minutes in the forward progress of your story, and start this sentence, "The reason this is important to you is..." and then finish it with a benefit to that specific audience. Or pose this rhetorical question, "What's in it for *you*?" and answer it with a benefit.

Only then can you make your call to action; only then can you make the sale.

Section II

Graphics: How to Design PowerPoint Slides Effectively

31

Vinod Khosla's Five-Second Rule

A Sanity Check for Every Presentation

Venture Capitalist Vinod Khosla, whose communication philosophy, "Message sent is not the same as message received," you read about in Chapter 19, "Vinod Khosla's Cardinal Rule," is a keen observer of *all* aspects of communication, including presentation graphics. During his 25 years in venture capital, Mr. Khosla has seen as many presentations as—if not more than—a presentation coach. Most of them were on Mondays, the day that Silicon Valley venture firms traditionally allocate to screening pitches from new companies. One by one, the start-ups stand up to be judged and then funded—or not. Once a candidate company makes it into the Khosla Ventures portfolio, Mr. Khosla continues to monitor and critique the presentations they develop to pitch to their potential customers and partners.

For all of them, he applies another cardinal standard: The five-second rule: Mr. Khosla puts a slide on a screen, removes it after five seconds, and then asks the viewer to describe the slide. A dense slide fails the test—and fails to provide the basic function of any visual: to aid the presentation. A simple slide makes its point instantly—fulfilling the name of the 150-year old company that produces calendars, planners, and organizers: AT-A-GLANCE.[1]

By applying his simple rule, Mr. Khosla addresses two of the most important elements in presentation graphics: Less is More, a plea all too often sounded by helpless audiences to hapless presenters; and more important, the human perception factor. Whenever an image appears on any screen, the eyes of every member of every audience move to the screen to process the new information, and they do so reflexively. The denser the image, the more time the audience needs to process. At that moment, they stop listening to the presenter. Nevertheless, most presenters continue speaking, further compounding the processing effort for the audience. As a result, they shut down.

Game over.

The simple solution to this pervasive problem is to follow the example of television news programs. Despite a vast array of sophisticated graphics capabilities at any broadcaster's disposal, all they ever show is an image composed of a picture and a few words. This image serves as a headline for the story that the anchorperson tells.

For presentations, consider yourself the anchorperson and design your slides to serve as the headline for your story. When your audience sees that simple an image, they understand it at a glance—Mr. Khosla's five-second rule—and they keep listening to you.

Game on!

32

Don't Raise the Bridge, Lower the Water

Better Box Thinking

The aphorism, "Don't raise the bridge, lower the water," has applications from the soaring heights of architectural design to the ordinary task of presentation graphics; the common denominator in both being the importance of thinking outside the box.

In a paper called "Thinking More Effectively about Deliberate Innovation," Christopher M. Barlow, PhD, a member of The Co-Creativity Institute,[1] said that the familiar phrase "forced me to a new perspective: creativity is not a change in the problem, it is a change in us, a change in our thinking that makes the already possible solutions obvious."[2]

Mr. Barlow identified the problem in the aphorism as how to "get the boats past the bridge," and then went on to say, "If I ask you to design a lift bridge, and you begin describing the building of a dam and lock to lower the water level, I have to wonder about your sanity or intelligence.... When some of the alternatives [are] made obvious by the new viewpoint are better than the best of the old ideas, we call it creativity." He summarized the creative process as "Not out of the box thinking, better box thinking!"

An example of better box thinking in the usually boxy world of architecture comes from the International Commerce Centre (ICC) in Hong Kong, a 108-story, 1,588-foot building, the fourth-tallest tower in the world.

Because tall buildings tend to sway in the wind, architects—like Paul Katz of Kohn Pedersen Fox Associates who designed the ICC—seek innovative ways to mitigate the risk. According to a *Wall Street Journal* story about the tower:

> *Most skyscrapers utilize pendulums or dampers, designed to transfer the motion of the building to mitigation devices.... Mr. Katz designed the entire ICC to provide wind buffers. Instead of meeting in corners, the sides join in recessed notches. Its scaled surface—which gives the building a dragon appearance beloved by Chinese—also breaks up wind force.*[3]

For presenters—who usually think within the strict confines of outbound corporate marketing boxes—better box thinking involves consideration of the audience. In most of today's unilateral and overloaded business presentations, thinking about the audience all too often goes missing in action.

One of my clients, let's call him Jason, is a marketing manager for a Silicon Valley telecommunications company, and he was assigned to develop his company's slide show for a new product launch. Jason created a network diagram slide in which all the labels were crammed into small boxes (pun intended), each box containing two- and three-line captions. Each second- and third-line in each box—which is known as word wrap—makes it more difficult for audiences to read than one-liners.

Word wrap forces the audience's eyes to make extra movements. Try this simple exercise: Find an existing slide that has word wrap in the bullets and/or the captions. Make a duplicate of that slide and then edit out all the instances of word wrap in the new slide. Then put both the original and new versions into Slide Show and click back and forth between them. Feel the difference in your eye movement?

When I suggested that Jason trim the labels to one-liners, he asked, "Should I make the text smaller or the boxes larger?"

I replied, "Don't raise the bridge, lower the water!"

Jason smiled in recognition that it was more important to make the slide easy for the audience to read than for him to create.

That's better box thinking.

33

Jon Stewart's Right

Positioned on Purpose?

No, not Jon Stewart's right as in "correct," and, given the liberal point of view of the host of Comedy Central's *The Daily Show*, certainly not as in "right wing." I'm referring to Jon Stewart's right side where he shows the video clips of people and events he satirizes or mocks. Is this positioning arbitrary or intentional?

Audiences in Western cultures read from left to right. Therefore, depending on the message you want to convey, you should design, animate, and display your presentation graphics so that they follow or fight that predisposition. Movement to the right creates positive perceptions, movement to the left negative.

In Microsoft PowerPoint animation, the left and right movements occur in two general options: between slides (Slide Transition) and within a slide (Custom Animation). Although the direction of movement is the same in each option, each has a different nomenclature. Movement to the right in Slide Transition is called "Wipe Right"; movement to the right in Custom Animation is called "Wipe from Left." Because your audiences' eyes are accustomed to the left-to-right movement, make your default animation follow that same natural movement.

Movement to the left in Slide Transition is called "Wipe Left"; movement to the left in Custom Animation is called "Wipe from Right." When you want to send a negative message such as the shortcomings of competing products, past problems your company has conquered, or market forces that pose major obstacles for your industry, use this counterintuitive effect.

Moreover, whenever you present, be sure that the screen on which you display your slide show—whether a large projection screen or a small laptop—is located to your *left* as you face your audience. This positioning creates that familiar left-to-right movement for your audience. Every time you click to a new slide, their eyes will travel from you to your image in a smooth, fluid movement. If you present with the screen to the right, every new slide will cause your audience to make a resistant move to the left that will force them to read backwards—a larger problem when your slide is composed of text.

Jon Stewart positions the images of the targets of his humor to his right, forcing his audiences to move to the left—with friction—to take in the images. Friction in the movement produces a fractious perception.

Is this positioning arbitrary or intentional? Is Jon Stewart sending us a message?

To use the words of one of his favorite targets, "You betcha!"

34

Misdirection

Magicians and Graphics

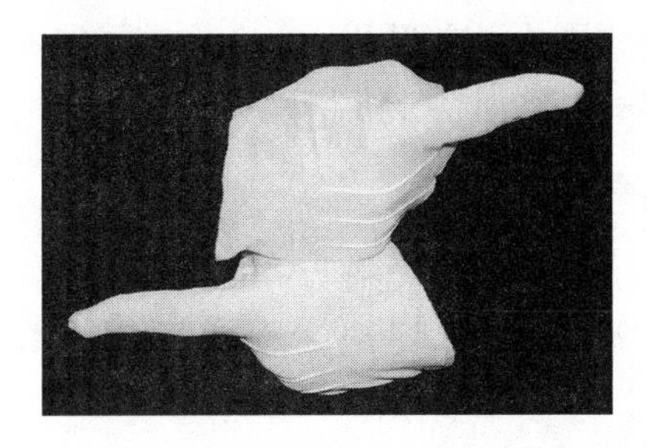

A documentary called *Make Believe,* about a group of teenagers who compete in a championship competition for magicians in Las Vegas, offers a lesson in how to display presentation graphics. The film focuses on one of the most fundamental techniques used in magic: misdirection, or getting the audience to look in one direction while the magician performs a trick in the other direction. Misdirection is based on the reflexive action of the audience's eyes—to look at new visual information involuntarily.

Presenters often inadvertently abuse this reflexive action of their audiences' eyes because of one of the most commonly held *false* beliefs about presentations: that if presenters turn to look at their slides, they will appear to be unsure of their own content. However, if a presenter does *not* turn to look at a new slide and continues to look at the audience, the audience will become conflicted. Their optic reflexes will force them look at the new image *involuntarily*. At the same time, the audience will also feel compelled to return the presenter's gaze. Driven by these two *opposing* impulses, the audience will become confused, and their eyes will rapidly shuttle back and forth between the screen and the presenter.

The difference between the false belief and neurological fact can be defined as B-School (for business school) thinking versus C-School (for cinema) thinking. B-School teaches students to demonstrate assuredness; C-School teaches students to understand human sensory perception. Cinematographers and film editors understand the powerful subconscious physiological and psychological forces that impact audiences. These professionals play to these dynamics; they shoot and edit sequences to create positive or negative feelings to express the action of the story. In presentations, you want to create only positive feelings in your audiences.

Therefore, as a presenter, the instant a new slide appears, you must turn to look at the screen. In fact, **turn to look at the screen with *every* click of *every* slide**. Every time you turn to the screen, your movement will lead your audience turn to look where you are looking. Both you and your audience will arrive at the identical point in your presentation, in synchronization.

In the *Wall Street Journal* review of *Make Believe,* film critic Joe Morgenstern wrote, "In magic, as distinct from filmmaking, misdirection is a good thing."[1] To which I add, in presentations, misdirection is a bad thing.

Every time you click, always turn to look at your screen.

35

Obama Makes a PowerPoint Point

The State of the Union and Presentations

Ever since George Washington's first State of the Union address in 1790, U.S. presidents have delivered their constitutionally mandated annual speeches without visual aids. President Obama's second State of the Union was no exception, but that particular speech offers a lesson in presentation graphics.

Just as all his modern predecessors have done, Mr. Obama spoke with the aid of only a TelePrompTer, but he did offer a dissenting opinion on the annual obligation.

In 2011, the ritual began, as they always do, with a grand ceremonial entrance into the House of Representatives chamber. Mr. Obama made his way through the packed chamber with all the pomp and circumstance of a coronation, shaking hands and/or hugging the members of congress and government who lined the crowded aisle.

As he approached the podium, he stopped to greet Secretary of State Hillary Clinton who smiled at him and said, "Good speech."

The President smiled back and replied, "I don't need to deliver it now; everybody saw it."[1]

Mr. Obama was referring to the fact that text of his speech had been leaked in advance. His remark was a sardonic commentary on an aspect of State of the Union speeches that has an analog with presentations: dual functionality.

This is the "Presentation-as-Document Syndrome," the pervasive business practice in which PowerPoint slides serve as both a visual aid and a send-ahead in what is known as a "pre-read." This practice serves neither purpose; it actually diminishes both. Moreover, the practice defeats the purpose of the actual presentation itself because providing material in advance telegraphs the presenter's content.

Use your PowerPoint *only* to illustrate your presentation. Resist requests to send your slides ahead. If your prospective audience insists—and audiences are like the proverbial 800-pound gorilla, they do whatever they want to do—offer to send them an executive summary, and do it as a document created in Microsoft Word.

Word and PowerPoint sit side-by-side in Microsoft Office but, like East and West, never the twain shall meet.

Dual functionality is only effective when you serve as the message and the messenger and do them both at the same time, in real time.

The President of the United States must deliver the message in person, and so must you.

36

Go in the Right Direction

A Presentation Lesson from Akira Kurosawa

During his long and distinguished career, the great Japanese filmmaker Akira Kurosawa pioneered many innovative cinematic techniques that are applicable to today's presentation graphics. One is Mr. Kurosawa's creative use of the Wipe, a filmic transition between scenes in which a new image slides across an existing image and replaces it—like a curtain being drawn across the screen.

In today's fast-cut action films, the Wipe has fallen out of favor, but the effect is very useful in presentations where fast cuts can be jarring to an audience. More about speeds in a moment, but first let's look at how Mr. Kurosawa used Wipes in his 1952 film, *Ikiru*.

Ikiru, which means "to live" in Japanese, is a story about a man dying of terminal cancer; it was inspired by *The Death of Ivan Ilyich*, a novel by Leo Tolstoy. Two more recent films, *Biutiful* and *Beginners*, deal with the same personal subject, but Mr. Kurosawa provided an extra dimension to his film by adding social commentary—and expressing his point of view with the Wipe effect.

The leading character in *Ikiru* is a career civil servant in post-World War II Japan where stultifying bureaucracy weighs heavily on a Japanese society trying to recover. To illustrate that situation, a group of mothers shows up at a government office to lodge a complaint about a sewage pond in their neighborhood, but the bureaucrats duck their responsibility by sending the mothers to another office, and then to another, and another, giving them the runaround.

Mr. Kurosawa depicts the runaround in a montage of 16 very short scenes, transitioning from one office to another with the Wipe effect. The first nine Wipes alternate left and right, but the last seven all move to the left. As you read in Chapter 33, "Jon Stewart's Right," because audiences read from left to right, you should design, animate, and display your presentation graphics so that—depending on the message you want to convey—your graphics follow or fight that predisposition. Movement to the right creates positive perceptions, movement to the left negative.

In *Ikiru,* the crescendo of leftward moves builds to create a negative perception of the bureaucrats. Film historian Stephen Prince, who provided the commentary track on the Criterion Collection version of the film, called the montage "an assembly which is basically a Rogues' Gallery of scoundrels."[1]

The lesson for presenters is: If you want to send a negative message to discuss your competition, for example, use the Wipe Left transition. But if you want to create a positive perception of your own company, use the Wipe Right.

A note about speed: In all versions of PowerPoint prior to 2010, the Wipe Right transition moves at a fast speed with a hard edge, creating that curtain-across-the-screen effect. In the 2010 version, the default for the Wipe Right transition moves at a slower speed with a soft edge, creating the effect of a dissolve, and slowing down the transition. This is not to say that you revert to the machine gun cutting that most of our movies use today; instead, use the Wipe Right as your preferred transition, but change the speed from the default of one second to a quarter of a second.

Give your audiences positive perceptions, not a Rogues' Gallery of scoundrels.

37

PowerPoint and Movie Stunts

Use Graphics to Create Continuity

At first glance, movie stunts would seem to have nothing to do with presentations, but an article about stunts written for Salon by Matt Zoller Seitz, a freelance film critic, provides a valuable lesson for presenters. Mr. Seitz noted that the latest cinema technologies, while creating imaginative and exciting action, have lost the important element of continuity. He wrote that the modern movie "seeks to excite viewers by keeping them perpetually unsettled with computer-enhanced images, fast cutting and a camera that never stands still." As a result, he claimed, the film denies "the viewer a fixed vantage point on what's happening to the characters."[1]

In contrast, Mr. Seitz cited a 100-year old silent film of a man jumping out of a burning hot air balloon into the Hudson River. Although the film itself is lost, the key shot lives on in a Topps bubble gum card. The point of Mr. Seitz's historic reference is that the image is "a sustained wide shot that showed the diver in relation to the balloon and the Hudson River," thus providing context for the action and for the viewer. If that scene were shot today, he added, "We'd more likely see a flurry of shots, only one of which showed us the big picture."

The operative words above are "in relation to." In today's films, computer animation and fast cutting move the story along so quickly audiences overlook or are unaware of the lack of context. In today's pitches, presenters hurriedly cobble together a set of their existing slides, giving their presentations a one-after-another sequencing, in which no slide has any relationship to the preceding or following slides—and therefore no continuity for the presenter or the audience. An audience might try to figure out what one slide has to do with another first, but after a short while they give up and turn their attention to their mobile devices.

One solution is for the presenter to make verbal links between slides; another is to create continuity in the slide design using a technique called Anticipation Space. In Figure 37.1, you see two boxes side-by-side, one filled and one empty—the empty box creates a sense of anticipation.

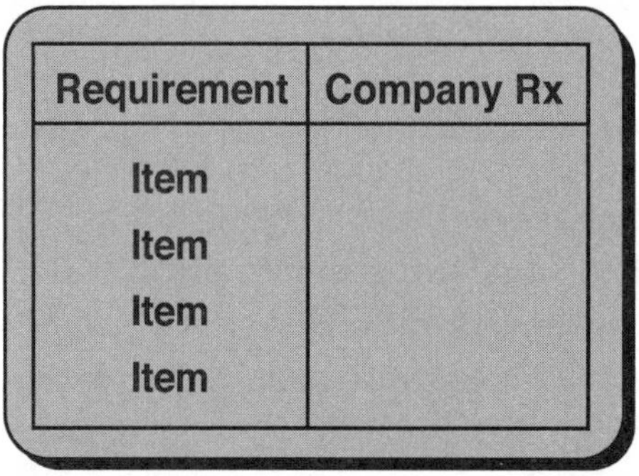

Requirement	Company Rx
Item	
Item	
Item	
Item	

Figure 37.1 Anticipation Space

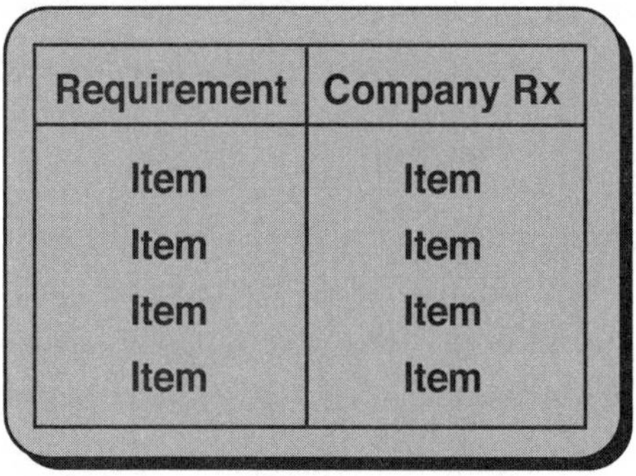

When the empty box is filled with a set of parallel items, as in Figure 37.2, it sends the message that your company's solution fulfills every requirement.

Requirement	Company Rx
Item	Item
Item	Item
Item	Item
Item	Item

Figure 37.2 Anticipation Fulfilled

Anticipation Space creates relationships, continuity, and much more: It makes your presentation easy for your audience to follow.

So easy, they might even look up from their mobile devices.

38

The Anti-PowerPoint Party

Another Precinct Heard From

A Swiss group calling themselves the Anti-PowerPoint Party launched their efforts in 2011, complete with a bright red octagonal STOP sign logo. In doing so, they took their place in a long line of detractors that stretches all the way back to 2003. The formal start of the criticism of the software was the *Wired* magazine publication of an article called "PowerPoint Is Evil: Power Corrupts, PowerPoint Corrupts Absolutely."[1]

The article was written by Edward R. Tufte, the noted graphics guru and professor emeritus of graphic design at Yale University. I've often challenged Mr. Tufte's opinions, but the critiques of PowerPoint go on unabated. My argument, simply and repeatedly stated, is to blame the penmanship, not the pen. A bad presentation is the fault of the user, not the tool.

To be fair, the Anti-PowerPoint Party does not fully advocate what its name implies. Its goal, as stated on its home page, is much more aligned with my argument:

> *We do not want to abolish PowerPoint®; we only want to abolish the PowerPoint®-CONSTRAINT.*
>
> *We want that the number of boring PowerPoint® presentations on the planet to decrease and the average presentation to become more exciting and more interesting.*[2]

Nevertheless, the hue and cry of the Anti-PowerPoint Party was echoed by Lucy Kellaway, who writes the excellent "Business Life" column for *Financial Times.* In her article about the Party, Ms. Kellaway advocated that "the APPP needs a terrorist faction, which would advocate cutting the wire in the middle of the table that connects the laptop to the projector. ... Better still would be to campaign for an outright ban."[3]

Even better still would be to campaign for a correction of user errors by banning the use of PowerPoint for anything but presentations (not send-aheads or leave-behinds) and to subordinate and restrict its use during presentations to support and/or illustrate the presenter's narrative.

Joining this approach was a letter to the editors of *Financial Times* in response to Ms. Kellaway's article. The letter was written by Michael Baldwin, a presentation coach in New York, who wrote:

> *In print cartoons, there is a dynamic relationship between the image and the caption that makes them—the good ones—both inseparable and unforgettable. With proper training, presenters can employ this same dynamic to produce memorable and convincing presentations.*[4]

Heed Mr. Baldwin's analagous advice or your presentation will become a literal cartoon.

39

Signage Versus Documents

Drive Your PowerPoint Home

American Apparel, Staples, Knoll Furniture, and Lufthansa Airlines all share a common denominator with the New York subway system: Each of these diverse organizations uses the same popular typeface in its signage: Helvetica. A book called *Helvetica and the New York City Subway System* describes why their decision makers chose the font style:

> *For years, the signs in the New York City subway system were a bewildering hodge-podge of lettering styles, sizes, shapes, materials, colors, and messages.... Efforts to untangle this visual mess began in the mid-1960s, when the city transit authority hired the design firm Unimark International to create a clear and consistent sign system. We can see the results today in the white-on-black signs throughout the subway system, displaying station names, directions, and instructions in crisp Helvetica.*[1]

Figure 39.1 shows the clean look of several New York City subway train letter and number identifiers in Helvetica font embedded in simple circles.

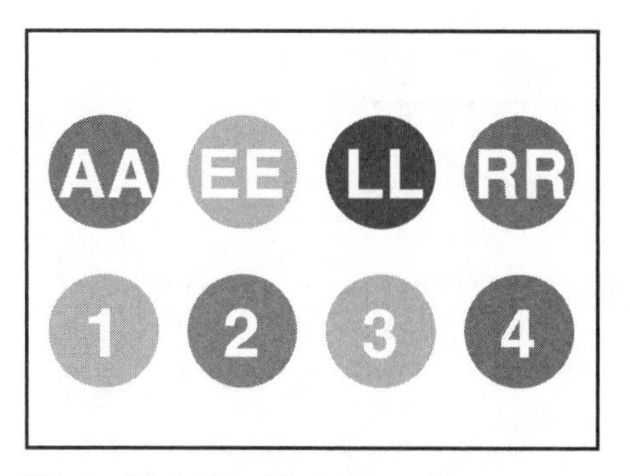

Figure 39.1 New York City subway train IDs

Helvetica, which is a sans serif font, is best suited for signage because its characteristic clean, straight strokes command attention. Sans serif is the font of choice in two universally familiar signs where attention is vital: **EXIT** and **STOP**.

Serif font, with its small decorative flourishes at the ends of the line strokes, is better suited for printed text because the little hooks help a reader's eyes to distinguish individual letters in long word strings. In printing, serif font is conventionally used for the body text. (Except for the body text on websites where, because of the low resolution of even the best computer screens, the hooks are difficult to read.) Just look at the body text of any article in the *New York Times* or the *Wall Street Journal*.

The point here for presenters is to draw an indelible boundary between documents that are meant to be read and graphics that are meant to illustrate. If you are creating a document, by all means use serif font.

But if you are creating presentation graphics, treat the text in your slides as signage or headlines. Design them in sans serif and compose them as headlines. Look again at any newspaper or magazine, and you'll see that the headlines are composed mainly of key words such as nouns, verbs, and modifiers, with very few articles, conjunctions, and prepositions; the latter are only necessary for complete sentences in the body text.

Unlike readers of text, your presentation audiences must process not only what you are showing but simultaneously what you are saying. If they have to process a large amount of sensory data, their eyes, ears, and, ultimately, their brains can go into overload—and they stop listening to you.

Instead, compose the text in your slides as headlines and do so in sans serif font—then provide the body text in your *verbal* narrative.

You are the presentation; your slides are the signage.

40

The Graphics Spectrum

Lives of Quiet Desperation

In 1845, the American author, philosopher, and naturalist Henry David Thoreau felt the need to get away from it all. He sequestered himself at an idyllic lake in the Berkshire Mountains for two years and wrote *Walden; Or, Life in the Woods* in which he famously observed, "The mass of men lead lives of quiet desperation."[1]

Mr. Thoreau's words are applicable to businesspeople today who lead lives of not-so-quiet desperation every time they have to make a presentation. Of all the many reasons for their desperation—time pressure, workload, and the fear of failure—perhaps the most pressing of all is the self-imposed practice of using their PowerPoint slides as not only the presentation graphics, but also as speaker notes, send-aheads, and leave-behinds. This multitasking approach produces images of exhaustive detail that serve none of the functions.

This bane of presenters has become a boon for another constituency: professional designers and authors who offer solutions to help businesspeople create simple, expressive, and purely illustrative graphical images. The best of the breed is Garr Reynold's marvelous book, *Presentation Zen,* which offers readers design concepts based on the principles of Japanese minimalism.

These polar opposites of the graphics spectrum leave a large underserved area in the middle consisting of presenters who want to break away from those exhaustive slides but find Mr. Reynold's Zen ideal too far a reach for them. At one end of the spectrum, some presenters protest, "But I'm not an artist!" (More about that in Chapter

43.) At the other end, others protest, "We don't have the time to do that!"

For that large majority, here is a simple set of guidelines for the two most basic types of garden variety graphics used in presentations today: bullet slides, as in Figure 40.1, and bar charts, in Figure 40.2.

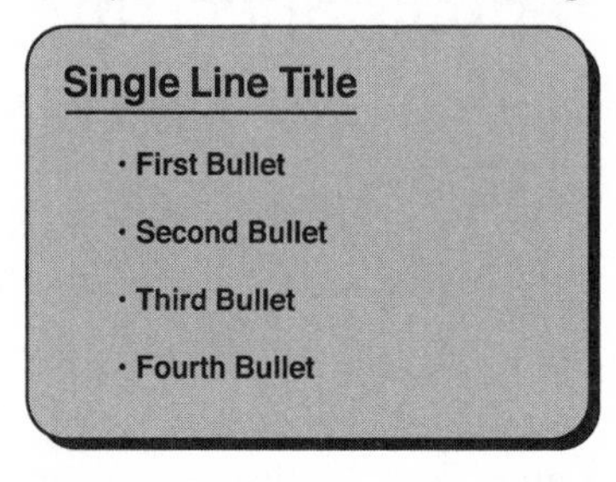

Figure 40.1 Effective bullet slide

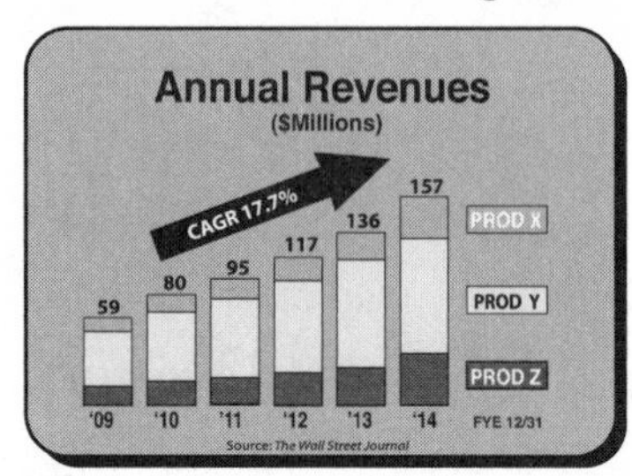

Figure 40.2 Effective bar chart

The guiding principles of this simple but effective bullet slide can be summarized in four bullets:

- Consider every line as a headline and not a sentence.
- Avoid word wrap by restricting every item to one line.
- Start each line with the same grammatical part of speech: verbs, nouns, and modifiers, etc.
- Keep all the elements related, creating pattern recognition.

The guiding principles of this simple but effective bar chart can be summarized in four bullets:

- Omit the Y-axis and place the numbers directly on the bars.
- Represent the legend in legible font size.
- Use large, color-coded labels.
- Make it easy for your audience by minimizing their search.

Or to paraphrase the last bullet in terms that Mr. Thoreau would appreciate, help your audience to lead lives free of desperation.

41

How Audiences See

Follow the Action

Arabole
ſalomõis

The large, illuminated letter at the upper left of this ancient text has been an established practice in documents ever since medieval times. It marks the beginning of a new document or a new section in a document. The practice has continued on into modern publishing where an enlarged first letter marks the beginnings of chapters in books and the beginnings of articles in magazines and newspapers. Now it becomes a factor in how we view computer—and presentation—screens.

EyeTrackShop, an eponymous Swedish start-up company, does exactly what its name says: Track eye movements to, as their slogan puts it, "identify where people look, for how long and in what order."[1] The company's technology uses webcams to follow and record how viewers' eyes scan images and then applies that information to help advertisers create effective ads and web designers to create effective web pages. By understanding the dynamics of how viewers scan ads and web pages, you can create effective graphics for your presentations.

One of EyeTrackShop's projects studied how users viewed the home pages of Facebook and Google+. The results, shown in Figure

41.1 and reported in the *Wall Street Journal*, found that in both cases, "Users' eyes head straight for the big status column in the middle of the screen, then over to the list of categories on the left side, then hop across to alerts on the right."[2]

Fixation Order

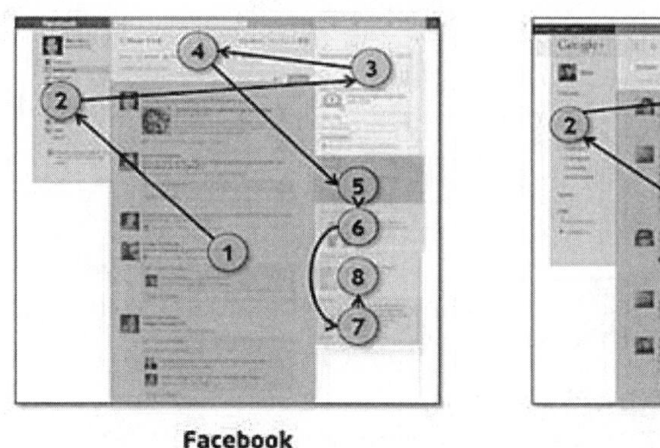

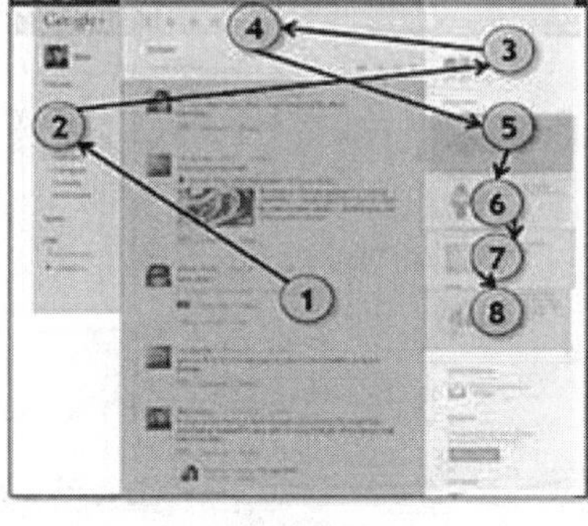

Figure 41.1 EyeTrackShop study of Facebook and Google+ Web pages

Those movements are driven by forces more powerful than the images on the Google and Facebook sites, two forces that drive the eyes of every human being:

- **Nurture.** In Western culture, because we have learned to read from left to right, our eyes start reading at the upper-left corner of documents.
- **Nature.** The optic nerves in all human eyes impel them to look at new images, and so, having started at the upper left, readers' eyes naturally—and involuntarily—move to the right to see the remainder of the new image.

As a result, human eyes do essentially what the eyes of the subjects in the EyeTrackShop study did: After centering on the full image, they move to the upper left to start reading, and then sweep across to the right to continue reading. In that same manner, whenever you click to a new slide, your audience's eyes start reading at the upper left of the projection screen and sweep across to the right.

If your slide is densely packed with images, numbers, and/or text, your audience's eyes will not see the entire image on the first move to the right. They will have to come back to the left and then go back to the right again. The denser the slide, the more times your audience's eyes will have to traverse the screen; the more traverses they make, the less they will hear of what you are saying.

Do you see where this is going? Back to the familiar Less is More principle, but now with this new added corollary: Reduce the number of moves your audience's eyes must make to understand your slide.

In the previous chapter, you saw these principles applied to the two most common slides in presentations today: bullets and bars. Please look at those two illustrations again and feel how your eyes naturally take in each slide: They start at the left and swing to the right.

Do the same for all your presentations. Design effective slides by reducing the number of eye moves your audiences must make.

Minimize the processing their eyes—and their brains—must do. Let them spend their energy and time focused on you.

42

Why Use PowerPoint at All?

A Picture Is Worth a Thousand Words

Considering the universal disapproval by critics, audiences, and even by presenters themselves, why would anyone use PowerPoint? The software—and its usage—have developed a universally negative reputation characterized by the common epithet, "Death by PowerPoint."

After all, the memorable speeches of history did not use PowerPoint:

- Cicero's orations in the Roman Forum
- Abraham Lincoln's Gettysburg Address
- Winston Churchill's World War II rally to arms
- Martin Luther King's Civil Rights speech
- John F. Kennedy's Inaugural Address
- Ronald Reagan's "Tear down this wall" Berlin speech
- Barack Obama's "Cinderella" keynote

So why, indeed, would anyone use PowerPoint? The simple answer lies in the aphorism, "A picture is worth a thousand words." Those familiar words are backed by a wide array of scientific evidence. One of the most thorough is an HP publication titled, "The Power of Visual Communication," which cites nine learned sources and concludes that:

Recent research supports the idea that visual communication can be more powerful than verbal communication, suggesting in many instances that people learn and retain information that is presented to them visually much better than that which is only provided verbally.[1]

Even more to the (Power) point is the opinion of Stephen M. Kosslyn, the author of *Clear and to the Point: 8 Psychological Principles for Compelling PowerPoint Presentations*, a popular book based on his work at the Department of Psychology at Harvard University. As Dr. Kosslyn put it in one of his academic studies:

The timeworn claim that a picture is worth a thousand words is generally well-supported by empirical evidence, suggesting that diagrams and other information graphics can enhance human cognitive capacities in a wide range of contexts and applications.[2]

The latest authority on visual expression is Hans Rosling, a Swedish medical doctor and statistician, whose revolutionary graphical display methodology electrified the high profile TED conference and made him an instant media and talk circuit rock star.[3] Ten thousand words would not be adequate to describe his technique, so please see for yourself in his video on YouTube.[4]

A *New York Times* article about Dr. Rosling described the impact of his graphics:

The goal of information visualization is not simply to represent millions of bits of data as illustrations. It is to prompt visceral comprehension, moments of insight that make viewers want to learn more.[5]

This is not to say that you should attempt to scale Dr. Rosling's heights of creativity, but to be inspired by his simple yet animated approach to depicting statistics. (You can download his specialized Trendalyzer software from his site for free.[6])

Be inspired even more by his "Five Hints for a Successful Bubble Presentation" that he offers with the download. Especially the second:

Explain what is shown on the vertical and horizontal axis by color and size of bubble before you start moving the bubble.[7]

In that one sentence, Dr. Rosling, his dynamic software notwithstanding, validates the primacy of the presenter over even his superb graphics. More than that, he tells you not only *why*, but *how* to use PowerPoint.

That still leaves open the matter of *what* to put in the PowerPoint—the subject of the next chapter.

43

"But, I'm not an artist!"

Rx: Infographics

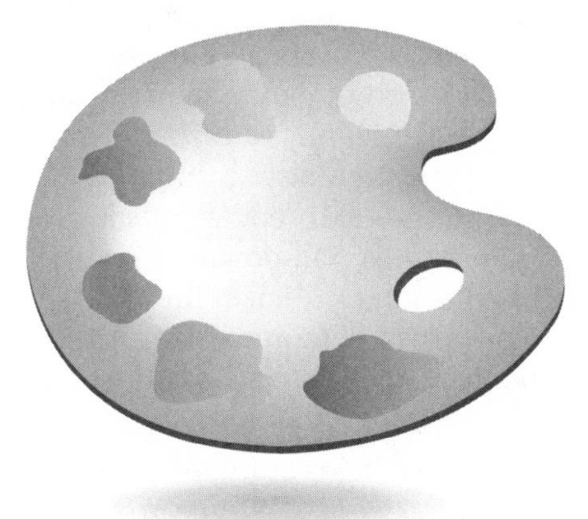

In my constant effort as a coach to persuade business-people to remember that a picture is worth a thousand words and to avoid the dreaded "Presentation-as-Document Syndrome," presenters often protest, "But I'm not an artist!"

Cast adrift from their familiar text slides, presenters are reluctant to try alternatives. However, you don't have to go out and buy a painter's smock and beret to break the mold of an endless parade of boring bullet slides.

Begin with overarching concept that the primary—and sole—purpose of your PowerPoint is to illustrate your narrative. Remember my often-repeated (because it *still* hasn't taken hold) recommendation that your business slide show should follow the example of television news broadcasts: the anchorperson tells the story and the graphics serve as a headline that captures the essence of the story.

Then design your presentation headlines as "infographics" or "data visualizations." Visual.ly, the world's largest community for creating and sharing infographics, defines these terms as follows:

...Infographics are images created to explain a particular idea or dataset. They often contain beautiful graphics to increase their appeal and help catch your attention. Many of them use data visualizations.

Data visualizations represent numerical data in a visual format. They can be anything from a simple bar chart to a complex three dimensional CAT Scan representation.[1]

But you can go beyond the usual charts and venture into more vivid images to communicate and illustrate your story. You have at your disposal a number of resources to convert text into images and to inspire your thinking visually:

- **Google and Bing.** Each of these powerful search engines has an "Images" feature. Just go to the search bar on either site, type in a keyword, and you'll see a broad array of photos, clip art, and line drawings. You can also search for synonyms of a keyword. For instance, "jail," "prison," and "penitentiary" bring up multiple variations of incarceration images. Moreover, as soon as you type in a keyword, each site offers a pull-down menu with other variations. "Jail" brings up "jail bars," "jail cell," and "jail house," and each of them brings up even more images—all in the interest of getting your creative juices flowing; to think outside the plain vanilla text box.

  Be aware, however, that many of the images on these sites may require payment for high resolution copies and/or royalties. Below is a list of ten websites where you can find free or low cost images.

- **Royalty-free/public domain image websites.**

  www.freepik.com
  www.images.google.com
  www.house.gov
  www.openclipart.org
  www.Picdrome.com
  www.publicphoto.org
  www.senate.gov
  www.TotallyFreeImages.com
  www.usa.gov/Topics/Graphics.shtml
  www.ushistoryimages.com

- **Visual.ly.** Visit this excellent graphic community site and see what their "data visualization enthusiasts" have created. Browse its pages and sample the many impressive infographics their members have posted. They will inspire you to think visually. The site also provides a tool to step you through the creation of your own infographic.
- **Microsoft PowerPoint.** The industry standard presentation software itself offers multiple ways to turn plain vanilla words into interesting graphics. Just click the Insert tab on the top ribbon, and another tab opens with the following graphical choices: Table, Pictures, Clip Art, Photos, Shapes, Charts, and SmartArt. The latter provides an almost infinite array of shapes, colors, and textures to enhance the look and feel of your text. Look at the difference that embedding text in simple shapes and shading can make with the identical text in Figure 43.1.

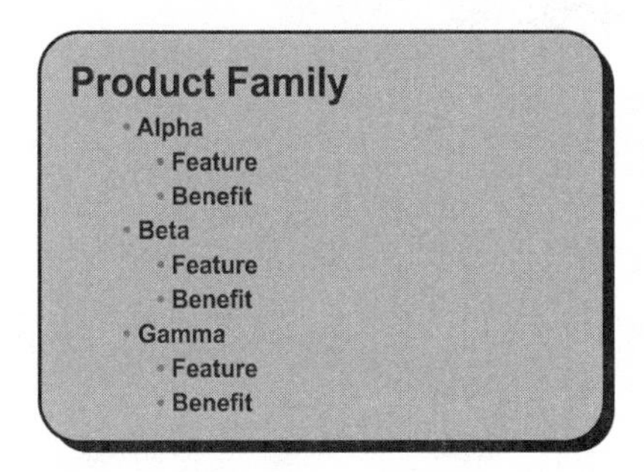

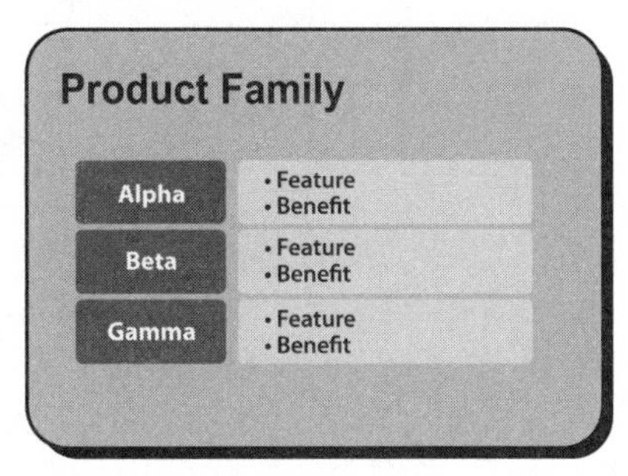

Figure 43.1 Add visual flavor to plain vanilla

Now, with your palette of four different options—Google/Bing, free images, Visual.ly, and Microsoft PowerPoint—are you ready for your artist's smock and beret?

44

The Kindness of Strangers

Stand and Deliver

Tennessee Williams's Pulitzer Prize-winning 1947 play *A Streetcar Named Desire* and its 1951 Oscar-winning film version starring Marlon Brando is an American classic. The story's characters and dialogue have become a familiar and recurring element in our cultural references. One of the most famous was a line spoken by one of the leading characters, Blanche Dubois, an aging beauty whose troubled life is going awry. At a pivotal moment in the story, Blanche says, "I have always depended on the kindness of strangers."

So famous is the line, many other authors have used *The Kindness of Strangers* as the title of their books, among them novels, travel books, and most appropriately, a 1997 biography of the line's original author, Tennessee Williams.

Most of these references focus on the second part of the quote and, in doing so, shift from the original context of Blanche's helplessness to the benevolence of others. In the highly competitive world of business, audiences occasionally—but only very occasionally—bestow benevolence on presenters.

One such person is Guillaume Estegassy, a Sales Strategy Manager at Microsoft Corporation, who said, "I have become so used to presenters who have terribly complicated slides and who do such a poor job of presenting them, I usually just stop looking at the slides and try to follow the story. I give them the benefit of the doubt."

Mr. Estegassy is a rarity. In this high-speed, broadband fiber optic world where most people have the attention span of a firecracker, presenters cannot be dependent on the kindness of their audiences. Instead, presenters must anticipate how their audiences will react to their slide show, not just in the design stage, but in preview.

Just as baseball and football teams play exhibition games, Broadway plays do out-of-town tryouts, and software applications do beta-tests, you should do a preliminary run-through your presentation—and do it with a mock audience composed of your colleagues. In most business practices, this seemingly obvious step is often bypassed in the rush to completion.

Take a lesson from Deborah Landau, who directs the creative-writing program at New York University. In an article in the *Wall Street Journal*'s "Word Craft" column, Ms. Landau wrote, "Like many writers, I'm often too close to my work to see it clearly, and a fresh pair of eyes can be invaluable. Kindness is a luxury I can't afford."[1]

Steve Ahlbom, the Creative Director of Artitudes Design, a Seattle-based graphic design company, knows the value of a fresh pair of eyes. (Full disclosure: I have worked extensively with Steve and Artitudes in the creation of our Power Presentations website, and have great respect for their skills and professionalism. So does Microsoft Corporation, which outsources many of its presentation graphics to Artitudes.)

In working on a project for Microsoft with Steve, I showed him how I coach participants in the Power Presentations program: They stand and deliver in real time, and I role play their real audience, the presentation equivalent of a pre-Broadway preview.

A light bulb went off for Steve. He said, "We never do that! We just design the slides for our clients and then send them off! I'm going have all our designers stand and deliver the slides themselves!"

Stand and deliver your presentation *several times* before your next important event—and you won't have to depend on the kindness of your audiences.

45

No More Mind-Numbing Number Slides

Five Easy Steps to Bring Your Presentation to Life

Businesspeople are perpetually faced with the challenge of one of life's greatest burdens: presenting number slides without numbing their audiences into a soporific stupor. This narcoleptic effect is the result of four common missteps perpetrated by most presenters:

1. The presenter starts each slide by saying, "Now I'd like to talk about..." forcing the audience to restart the presentation with each slide.
2. The presenter reads the words on the slide verbatim, causing the audience to feel patronized and become resentful, thinking "I can read it myself!" (Just remember that the first time anyone read to you was to put you to sleep, and thus you are forever programmed.)
3. The presenter discusses the general subject of the slide without referencing the specifics on the slide, splitting what the audience sees and what they hear, forcing them to dart back and forth between the screen and the presenter, causing complete confusion.
4. The presenter recites only the data on the slides, adding no value.

Therefore, the problem is in the presenter's narration more than in the design of the slide itself. Of course, it is important to wield a sharp razor in the graphic design, slashing and trashing extraneous data, Keeping It Simple, Stupid. But even the most minimal design must be accompanied by a clear and consistent narrative.

Here then is a simple solution for each of the four common errors, one for each error, plus one bonus solution, linking your slides into a fluid story narrative.

1. **Title *Plus*.** To avoid the restart effect, start each slide with a Title *Plus*, a single statement that captures the overview of the entire image by referencing the title of the slide *plus* the other elements on the slide. For a financial slide with 5 bars, say, "Here you see our revenues for the past five years." Or for a financial slide with 20 bars, say, "Here are those same annual revenues in quarterly increments." For a pie chart say, "This slide represents the percentage of our revenues by region."

   You can also use the Title *Plus* to describe other than number slides. For a bullet slide say, "These are the four steps we intend to take on our path to profitability." For a complex technology diagram say, "This is our comprehensive technology architecture."

2. **Paraphrase.** To avoid the verbatim reading effect, paraphrase or juxtapose the words on the slide, or use synonyms. For instance, if the slide title reads *Significant Revenue Growth*, say, "Our revenues have grown impressively." Or if the slide title reads *Multiple Market Drivers*, say, "These are the many forces driving our market." If the slide title reads *Broad Patent Portfolio*, say, "We have strong Intellectual Property protection." Your audience can easily make the interpolation—and not feel patronized.

3. **Navigate.** To avoid the split perception effect, describe the images on the slide by navigating the audience's attention with your spoken words. For a pie chart, say, "The largest wedge is the green with 55 percent, moving clockwise, the middle wedge, in yellow, is 38 percent, and the smallest, in blue, is 7 percent." For a table, say, "The Y-axis represents speed from low up to high, and the X-axis represents costs from low out to high."

   In addition to making it easy for your audience to follow and understand, this navigation technique has an extra benefit: It displaces the ubiquitous pointer. For some inconceivable reason, pointers, whether the retractable fixed type or the frenetic red dot laser model, have become standard equipment in presentation environments around the world. Presenters then pick up the pointers and brandish them as threatening weapons.

Verbal navigation is user- and audience-friendly.

4. **Add value.** To avoid the recitation trap, add value, dimension, and depth to your narrative by referencing information that does *not* appear on the slide, such as data, examples, quotations, or analogies. Add brief discussions to add value.

   Financial prospectuses have a boilerplate section called "Management's Discussion and Analysis." Make this the theme for every presentation, especially number slides. Discuss and analyze, don't recite.

5. **Bonus: Linking words.** You can create continuity from slide-to-slide with a technique that skilled writers use to create continuity in their narratives. They chose a word or a phrase from one paragraph and repeat that word or phrase in the subsequent paragraph, thus connecting the two paragraphs. The same technique can be applied to two consecutive slides. Let's say you have one titled *Significant Revenue Growth* and another titled *Margin Improvement*. When you click to the margin slide, say, "Our impressive *revenue growth* has helped us improve our *margins*." Or if one slide is titled *Broad Product Line* and the next is titled *Leading Market Share*. When you click to the market share slide, say, "Our state-of-the-art *products* have made us the *market leader*."

   Contrast this linking technique with the conventional rote transition that maddeningly starts each slide, "Now I'd like to..." which provides no link at all.

The linking words technique, along with the other four solutions, brings logic and continuity to what is essentially a disparate and interchangeable laundry list of data. It also brings vitality to your number slides—as well as to *all* your slides—and to your audience.

Section III

Delivery Skills: Actions Speak Louder than Words

46

Eight Presentations a Day

Cause and Effect

For each of the past 40 years, the American Electronics Association, under the auspices of TechAmerica, holds its annual Classic Financial Conference at which more than 1,800 technology companies and more than 6,000 investors come together in one venue. Executives from participating companies deliver eight to ten iterations of their corporate pitches in one day, giving existing investors multiple opportunities to get updates on their holdings, and potential investors multiple opportunities to learn about new businesses.[1]

Noland Granberry, the Chief Financial Officer of public company Silicon Image, Inc., a leading provider of advanced, interoperable connectivity solutions for consumer electronics, mobile, and PC markets, presented at the one of the conferences. Over the course of his eight presentations, Mr. Granberry experienced a progressive dynamic that provides a helpful lesson for any presenter.

As he delivered each iteration of the Silicon Image pitch, Mr. Granberry felt his comfort zone increase, resulting in what he believed to be a smoother delivery; he also felt he portrayed his company more confidently and authoritatively. This progressive evolution also produced another benefit.

At the end of each presentation at the AEA conference, the presenters open the floor to questions. As a matter of course, most investors ask challenging questions—after all, they want to be sure that their investment is in good hands. Mr. Granberry got his share of challenging questions in his sessions, but he noticed that, as they day wore on, the questions became less challenging—in direct proportion to the improvement in his presentation delivery.

Think about that: Mr. Granberry delivered the identical content each of the eight times and nothing changed but his delivery. He had experienced—in real time—the power of Verbalization, a rehearsal method of speaking the presentation aloud multiple times.

Unfortunately, most presenters bypass *any* rehearsal at all. If they do rehearse, they either mumble, "Blah, blah, blah..." or they talk *about* their presentation, "On this slide, I'll discuss...." Neither of these methods is Verbalization. Verbalization means delivering the presentation in rehearsal as if there were an actual audience in the room, "Good morning. Thank you for giving me the opportunity to...."

This powerful—and yet underutilized—practice technique clarifies the words and the flow of the content, giving presenters the assurance to present with conviction. If you prepare, practice, and deliver your presentation thoroughly, you can not only present your story more effectively, you can also diminish the challenges in your Q&A session.

Verbalization worked for Noland Granberry; it can work for you. You can control your own destiny.

47

Sounds of Silence

Presentation Advice from Composers and Musicians

Stephen Sondheim, the legendary Broadway composer and lyricist, whose countless Broadway musicals have earned him eight Tony Awards, multiple Grammy Awards, and a Pulitzer Prize, decided to culminate his career by sharing the secrets of his craft in two impressive books:

- *Finishing the Hat: Collected Lyrics (1954-1981) with Attendant Comments, Principles, Heresies, Grudges, Whines and Anecdotes*
- *Look, I Made a Hat: Collected Lyrics (1981-2011) with Attendant Comments, Amplifications, Dogmas, Harangues, Digressions, Anecdotes and Miscellany*

When the first book was published, the *New York Times* called upon an equally legendary songwriter, Paul Simon, the former partner of Art Garfunkel, to write a review. The book and the review contain, valuable advice about words and music—shared by Mr. Simon and Mr. Sondheim—that is applicable to presentations: how audiences process what they hear.

During a discussion of rhyming lyrics in the book, Mr. Sondheim described the consequence of a poor rhyme:

> *By the time the ear has figured out what is actually being sung, the singer is in the middle of the next line and the listener has to waste his concentration on catching up.*[1]

In the review, Mr. Simon added his own view on the subject:

I have a similar thought regarding attention span and a listener's need for time to digest a complicated line or visualize an unusual image. I try to leave a space after a difficult line—either silence or a lyrical cliché that gives the ear a chance to "catch up" with the song before the next thought arrives and the listener is lost.[2]

Given his famous 1965 song, "Sounds of Silence," Mr. Simon knows whereof he speaks.

The two composers' opinions are echoed by those of two jazz musicians. Trumpeter Dizzy Gillespie, the father of BeBop once said, "It's taken me most of my life to know which notes *not* to play." And tenor saxophonist Houston Person, in a *Wall Street Journal* profile, said:

Silence is as much a part of the music as the notes are. After all, if you were to speak to someone and not pause here and there everything would have equal importance. You use silence to underline something, whether you play it on an instrument or speak it in a conversation.[3]

Once is an opinion, twice an affirmation, three times a trend, and the fourth makes it sound advice you would do well to follow in presentations.

Put spaces between your words—and images—to allow your audiences to absorb the points of your message. When you present, speak with a cadence that includes pauses between your phrases, and when you design your slides, surround your images with empty space.

Listen to the masters.

Pause.

48

Stage Fright

A Close Cousin of Writer's Block

In Chapter 12, "Writer's Block," you read about how the hero of the Hollywood film *Limitless* cured his writer's block with a drug that stimulated his creative capabilities. Concurrent with the film's opening, a related article about creative paralysis appeared in the *New Yorker* magazine. Staff writer Dana Goodyear profiled Barry Michels, a real life therapist who treats blocked Hollywood screenwriters with his own unique methodology derived from the concepts of Jungian psychology.

Mr. Michels, whose starting rate is $365 an hour, also treats the stage fright that movie colony writers and other creative people face when they have to pitch their ideas—a subject near and dear to the solar plexus of every presenter. The presentation equivalent of stage fright is the pervasive fear of public speaking. Although Hollywood pitch meetings are anything but public, and Los Angeles is 3,000 miles and a galaxy away from Wall Street, the angst is just as real and just as pervasive.

Mr. Michels works in tandem with his mentor, psychiatrist Phil Stutz. They treat their clients with three techniques that they call:

- Visualization
- The Shadow
- Dust

Ms. Goodyear described how Mr. Michels uses Visualization:

Patients are told to visualize things going horribly wrong, a strategy of "pre-disappointment"...[that] involves imagining yourself falling backward into the sun, saying "I am willing to lose everything" as you are consumed in a giant fireball, after which, transformed into a sunbeam, you profess, "I am infinite."[1]

Mr. Michels' version of visualization is a 180 degree reverse of "guided imagery," a technique used by mental health professionals to get their patients to think *positive* thoughts and direct their minds toward a relaxed or desired state.[2]

Positive visualization is also used in sports where athletes envision successful outcomes: The racer crossing the finish line, the basketball going through the hoop, or the tennis ball landing in the perfect spot across the net.[3] This technique took wing in the 40-year old bestseller *The Inner Game of Tennis,* in which author W. Timothy Gallwey wrote, "Concentration is the act of focusing one's attention. As the mind is allowed to focus on a single object, it stills."[4]

Here is how Ms. Goodyear described Mr. Michels' second method, the Shadow:

...the occult aspect of the personality that Jung defined as "the sum of all those unpleasant qualities we like to hide, together with the insufficiently developed functions and the contents of the personal unconscious."[5]

The Shadow, like Visualization, is another negative point of view, as is Mr. Michels' third concept, Dust, which Ms. Goodyear said:

...involves pretending that your audience is covered from head to toe in dust—"a nice, thick, two-inch coat of dust, like you're going up into an attic and everything is covered, it's been up there for eight months."[6]

If Dust sounds familiar, it is. Mr. Michels and Mr. Stutz have coined a variation of the equally-senseless—and tasteless—pervasive recommendation that presenters imagine their audience, and job applicants imagine their prospective employer, naked.

If you're beginning to see an unorthodox trend here, you're not alone. In what has to be the understatement of the year, Ms. Goodyear observed, "Needless to say, neither therapist relates much to the wider analytic community, and both suspect that the techniques would be met with consternation."

The techniques also drew consternation from *Time Magazine* humor columnist Joel Stein who had a session with Mr. Stutz to discuss his panic over public speaking and concluded, "Phil has invented the only therapy technique I've ever heard of in which you leave with bigger problems than you walked in with."[7]

Mr. Michels' and Mr. Stutz's techniques would undoubtedly draw consternation from the business community for one simple reason: They ask their end users to apply imaginary solutions to real challenges. Businesspeople require specificity, the "Show me" principle.

To overcome the fear of public speaking, presenters should focus on the tangible results of their efforts: how the audience is reacting to their presentation in real time. When you present, if you see nodding heads, you can move forward with your story, but if you see furrowed brows or perplexed looks, you must stop and adjust your content to clarify or explain what you have just said. This simple act will produce head nods, and this immediate visible reaction will diminish the fear of failure that caused the stage fright in the first place.

In presentations, the endgame is a sea of nodding heads, not an image of the sun or a shadow or of a coat of dust. The only imaginary images are those of a bank of bright light bulbs going off over those bobbing heads, accompanied by a chorus of resounding "*AHAs*!"

See the cause and effect a change. "Show me the money!"

49

Swimming Lessons and Presentations

Deconstruct and Reconstruct

To teach swimming, coaches take novices through the component skills in progressive stages. The first lesson takes place out of the water, at the side of the pool, where the novice learns the arm stroke and the leg stroke separately. Then the novice gets into the shallow end of the pool and practices the arm strokes and leg strokes, still separately, but now with training equipment—flotation devices, kickboards, and the rungs of a pool ladder—to develop the skills further. As the training progresses, the novice puts aside the equipment and swims, first in the shallow end of the pool, then the deep end, and finally, in a lake or ocean. The fundamental aspect of this approach is to deconstruct the basics and then to reconstruct them progressively. It works in swimming—as it does in all sports—and it works in presentations.

In the Power Presentations program, we teach delivery skills in progressive stages, in the same manner that swimming coaches teach swimming, an approach that produces an unexpected benefit for presenters in the development of their stories.

In the delivery skills session of the program, we begin the deconstruction by excluding slides. The initial impetus for this approach was twofold:

- To reduce the complexity of the exercises (like the swimming lesson on dry land)
- To reinforce the primacy of the presenter over the slide show

Just as swimming lessons involve multiple repetitions, so, too, do we ask our participants to repeat a short pitch several times. The unexpected benefit is that, as presenters develop their delivery skills, they also improve their narrative flow. The reason this occurs is also twofold:

- Verbalization, a practice technique you read about in Chapter 46, "Eight Presentations a Day," creates familiarity and therefore fluency with the content.
- When presenters are freed from having to relate to their slides, they can concentrate more on their stories.

Slides, as they are conventionally designed today, hinder rather than help presenters. The all-too-common complexity of the slides forces presenters either to skim over them or, in the worst case, read them verbatim. This unholy alliance also fragments the narrative because each slide is discussed individually with no relationship to the next. When we exclude the slides in the practice, presenters focus on telling their story, connecting the dots, and creating a clear progression.

This is not to recommend that you eliminate slides completely. Given the deep entrenchment of PowerPoint in business today, that would be foolhardy. Instead, use simply designed slides that serve *only* to support your narrative.

In our coaching sessions, we ultimately bring the slides back into our exercises, the equivalent of putting the novice swimmer into the water. Then, having developed each component of each skill separately, a presenter can integrate all of them into a Power Presentation.

Here's how you can use the deconstruction/reconstruction process to develop your presentation:

- First develop a clear and logical narrative.
- Then design slides to illustrate your narrative.

- Combine your story and slides in a verbal run-through of your presentation, but do it seated in front of your computer screen so that you don't have to think about your eye contact, gestures, posture, or voice.
- Do another run-through standing up in a vacant conference room, presenting to the empty chairs as if they were occupied by audience members. This time, focus on your eye contact, gestures, posture, and voice.
- Do a dress rehearsal to colleagues or friends, integrating all the components.
- Dive into the water and swim like a fish.

50

Valley Girl Talk

Invisible Question Marks

One of the most familiar quotes from the Bible is, "When I was a child, I spake as a child, I understood as a child, I thought as a child: but when I became a man, I put away childish things."[1]

Unfortunately, legions of men—and women—have not followed the Biblical progression: They have become adults, but they still speak as children. They punctuate and malign their speech with repeated insertions of "like" and "you know?" But their most egregious and pervasive quirk is the sing-song pattern of their childhood. They speak their declarative sentences with rising inflection at the ends, forming questions rather than statements.

The effect is known as "up talking" or "Valley Girl Talk."[2]

Figure 50.1 Poet Taylor Mali

Taylor Mali, a spoken-word performer, voiceover artist, and poet captured this juvenile speech pattern in a clever poem called "Totally like whatever, you know?" This is the first stanza:

In case you hadn't noticed, it has somehow become uncool to sound like you know what you're talking about? Or believe strongly in what you're saying? Invisible question marks and parenthetical (you know?)s have been attaching themselves to the ends of our sentences? Even when those sentences aren't, like, questions? You know?[3]

Mr. Mali's poem was turned into an equally clever video called "Typography" by Ronnie Bruce in which the visual artist accompanies Mr. Mali's words with animated fonts. And Mr. Mali's voice reciting the poem on the soundtrack gives perfect expression to the rising inflection he disdains by ending his sentences with those "Invisible question marks."[4]

The remedy for Valley Girl Talk is to drop the voice at the end of sentences—in spoken language, at the ends of phrases—thus parsing the logic of the phrases. Dropping the voice to punctuate the phrases creates a crisp, clear, and *adult* cadence.

Cadence in speech is like rhythm in music. Think of the main theme of Beethoven's great Fifth Symphony and its famous pattern of three short notes followed by a long one:

Bam-Bam-Bam BAM.

From the sublime of Beethoven to the mundane, think of the universally familiar "Shave and a haircut...two bits." This rhythmic snippet is often expressed without words, as a knock on a door composed of five short notes followed by two long ones:

Bam-Bam-Bam-Bam-bam, BAM, BAM.

Try rapping your knuckles on your desk with just the five short notes...

Bam-Bam-Bam-Bam-bam...

It sounds incomplete, doesn't it? The final two raps resolve the musical phrase, just as dropping the voice in speech resolves the spoken phrase.

Readers of *The Power Presenter* will recognize this skill as "Complete the Arc;" the arc is the logic of the phrase, and the completion is the falling inflection that adds the *BAM, BAM* to your words—and puts away childish things.

51

"What do I do with my hands?"

A Simple Approach to Gesturing

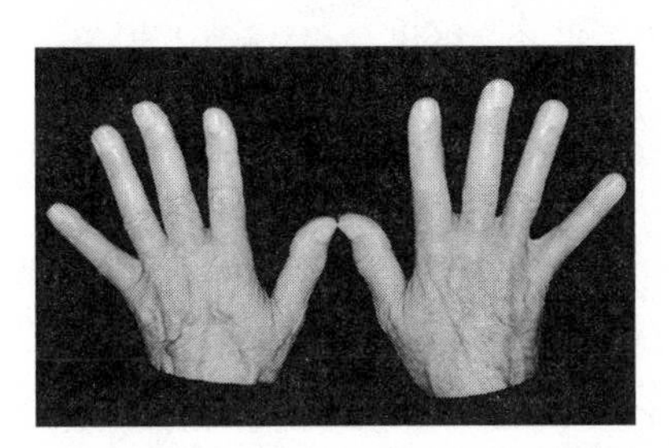

For the more than two decades I have been a presentation coach, the most frequently asked question (until now, but more about that in Chapter 54, "Rx: CrackBerry Addiction") has been, "What do I do with my hands?"

The answer is to use your hands to illustrate your words. But that seemingly simple instruction can lead to overkill if a coach or a presenter attempts to choreograph gestures. Choreography makes presenters feel like performers and worse, overloads them, as if asking them to pat their heads and rub their tummies at the same time.

An apocryphal story about Richard Nixon (who, after his historic presidential debate with John F. Kennedy, was forever identified as a notoriously stiff presenter) has it that he was coached to choreograph his gestures by writing prompts in the margin of his speech text. But, the story goes, Mr. Nixon lagged between his words and his gestures. He would say "two" and hold up two fingers a beat later, appearing even more awkward.

Choreography also runs counter to the admonition we've all heard ever since childhood from our mothers and teachers, "Don't speak with your hands!"

However, speaking with our hands has been a part of human communication ever since our cave dwelling ancestors—and even before that. A *Wall Street Journal* article by Matt Ridley reported that:

> *...our prehuman ancestors had only a modest vocabulary of shouts, screams and whines, but a richer and subtler vocabulary of gestures, shrugs and frowns. According to the primatologist Frans de Waal, apes, especially bonobos, use gestures more freely and flexibly than voice.*
>
> *According to another primatologist, Richard Byrne, gorillas have a large repertoire of gestures used to express specific meanings in the wild. And it appears that chimps learn a vocabulary of signs more easily than a vocabulary of sounds.*[1]

So with all due respect to mothers and teachers, you would do well to incorporate gestures to help you "express specific meanings." But how do you do that without feeling as if you are performing unnatural acts?

The answer is to let your hands do what comes naturally but not as choreography.

However, I will recommend one particular gesture to incorporate into your natural repertory—and if I recommend one, it had better have a very good reason.

Whenever you step up to the front of the room to present, you create a gap between you and your target audience. As a communicator your role is to close that gap. Therefore, do as the famous AT&T slogan recommends: *Reach out*.

When you reach out, you replicate the handshake, the universal symbol of human communication. The handshake is thought to have begun as a social custom during the Middle Ages. When the right hand—which was used to grip a sword or a dagger—was extended and empty, it indicated that the person was not armed. An open hand signaled, "I come in peace." Half a millennium of practice has instilled that same message in our modern culture.

So when do you reach out? In *The Power Presenter*, I recommend one overarching principle for delivery skills: that every presentation is a series of person-to-person conversations.

Therefore, whenever you present, have a conversation with each individual—each "you"—in your audience. And when you speak to a person, say "you" and, every time you say *"you,"* accompany the word with your hand and arm extended. *Reach out*.

- "Let me show *you*..."
- "Why am I telling *you* this?"
- "Do *you* see what I mean?"
- "What's in it for *you*?"

Your Mom would consider that quite polite.

52

"Look, Ma, no hands!"

Anchorperson or Weatherperson

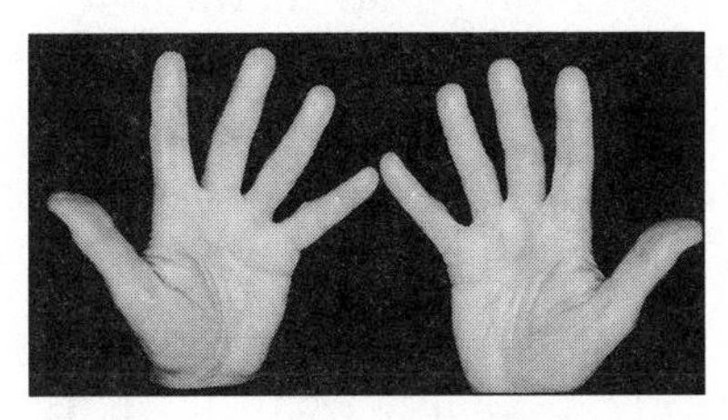

The advice about gestures in the previous chapter—to use your hands and arms to illustrate your words and to reach out to your audience, replicating a handshake—has an opposite point of view, and it comes from none other than The Great Communicator, Ronald Reagan.

Throughout his political career, Mr. Reagan rarely used any gestures. A DVD called *Ronald Reagan: The Great Communicator* is composed of clips from more than 100 public appearances during his eight years as president. In all the clips, he made an expansive gesture with his hands and arms only *once*.[1] In 1964, in one of his first appearances on the national political stage, Mr. Reagan gave a 30-minute speech to nominate Barry Goldwater and raised his arms only *once*.[2]

Mr. Reagan actually began to use this style during his formative years as a presenter. Between the twilight of his days as an actor and the start of his political career as the Governor of California, he spent eight years as a spokesman for General Electric Corporation. This role provided many opportunities to present in many venues.

One of them was as the host of *GE Theater*, an anthology series of television dramas. In one 1954 episode, he delivered his introduction standing, framed by stage lights, in front a blank wall of a movie studio. Attired in a smartly tailored tweed coat sprouting a natty pocket kerchief, he had his right arm propped on a stage light and his left hand in his trouser pocket. During the entire introduction, neither arm ever budged.

Figure 52.1 Ronald Reagan as host of GE Theatre

You might call this the "Look, Ma, no hands!" approach. The style worked—wonders—for Mr. Reagan. Would it work for you? The answer, as always, is to do what comes naturally to *you*.

An unnatural approach is to treat gesturing as performing. One variation of performing is to separate the use of hands into two camps known as "Anchorperson or Weatherperson." As we all know from television news programs, anchorpersons sit stock still at a desk, rarely using their hands, while weatherpersons wave their hands and arms about broadly to indicate weather patterns on a map. This division parallels the Ronald Reagan no-hands style vis-à-vis the gesture-to-illustrate style, but it does so in the context of performance.

If you are reading this book, it is highly unlikely that you are a performer or that you were auditioned for your position to evaluate your acting skills. You were undoubtedly hired on the basis of your background and the personality you presented during your interview, and that personality was *your* natural style.

For your gestures, heed the advice of Irving Berlin's song in the classic musical, *Annie Get Your Gun,* "Doin' What Comes Natur'lly."[3]

53

Foreign Films

The Pause That Refreshes

After several high school and college courses in French and Italian, a few classes at Berlitz, and numerous trips to France and Italy, I have developed enough facility in their languages to get by in their restaurants, hotels, and shops, but not nearly enough to have full conversations. However, I have also developed a taste for French and Italian cinema, and so my Netflix queue is populated primarily by such films. Of course, when I watch them, I have to rely on the subtitles for translation and drop my eyes to the bottom of the screen every time they change. As I do, my ears pick out some of the spoken words, but because the actors are natives they speak too quickly for me to follow them—except for the words at the ends of their sentences.

Therein lies a lesson for presenters.

Whenever actors, public speakers, clergy, or people in conversation end a sentence or a phrase, they usually pause. The pause gives listeners—the audience—time to absorb the words. But when a presenter stands up in front of an audience, the stress of the situation triggers an adrenaline rush, which produces time warp, which causes the presenter to speak faster and rush past the pauses.

Watch any Woody Allen film and you'll see the effect of stress on speech tempo. Most of his characters—as reflections of his own public persona—are neurotic people who get into complicated situations. As soon as the plot thickens, the characters' words accelerate like a Ferrari on the open road. This is amusing in a Woody Allen film, but it can damage a presentation because the rapid pace not only makes a presenter appear harried, it garbles the presenter's words. The latter problem is heightened when—in our globalized world—presenters speak English as their second language, or to audiences for whom English is a second language.

This is where we come full circle to the lesson from foreign films. Professional actors pay as much attention to the cadence of their speech as they do to the tone of their voices, and so, when talented actors end their sentences, they pause to punctuate the meaning of an idea. Presenters are not actors, but their ideas do fall into logical phrases. Presenters would do well to give their audiences—whether native English speakers or English-as-a-second-language speakers—a moment to absorb their information by pausing at the ends of their phrases.

I attended a presentation given by Xavier Martin, a Vice-President at Alcatel-Lucent, the global telecommunications equipment provider. Mr. Martin is French, and he started his pitch as fast as a racehorse bolting out of the gate. In the first moments, I heard him say "*zee ontairpreez,*" and didn't understand his words. But later on in the presentation, when he settled down and began pausing (if nothing else than to breathe) he spoke the words again. Only then did I realize that he had said, "the enterprise."

Take a lesson from foreign films and from the classic Coca-Cola slogan: Take "The pause that refreshes."

54

Rx: CrackBerry Addiction

Control Yourself!

As you read in Chapter 51, "What do I do with my hands?" has been the question most frequently asked of me as a presentation coach—until now.

Of late, the question has been, "How do you deal with audiences who are glued to their mobile devices?" The question, asked by distraught presenters, refers to a chronic malady known as "CrackBerry Addiction." Compounding the problem, those very same presenters, when they become audience members themselves, proceed to exhibit the same severe symptoms of the disease. The addiction is at epidemic proportions.

Quite frankly, I'm stumped for an answer. I've tried every technique I know—pregnant pauses, steely stares, provocative questions, innocent questions, polite requests, forceful demands, gentle nudges, outright pleas, periodic breaks, and even making a demonstrative point of shutting down my own mobile device—to no avail. The addiction persists.

Therefore, I'll approach the problem from a different angle; instead of trying to help presenters, I'll cast a wider net by recommending how *anyone* can escape the hypnotic allure of those glowing LCD screens. Admit it, you know that you are hooked, too. My hope is that if I can help move the needle only slightly, clearer minds might become more attentive audiences.

Professional writers, for whom concentration is critical, are often derailed by the double-edged sword of the Internet: They use it to find material, but they often go off on search sidetracks that interrupt their creative process. In an article for the *New York Times* Book Review, travel writer Tony Perrottet described one of his lengthy web detours and added that he is not alone in literary circles, "everyone I know acknowledges the problem of digital distraction." Mr. Perrottet went on to note that some writers "have made gestures toward enforced self-denial," and gave the example of author Jonathan Franzen who wrote his bestselling novel *The Corrections* "in a dark room wearing earplugs, earmuffs and a blindfold, and confessed to blocking his Ethernet port with Super Glue."[1]

Extremes measures, but not as extreme as those of Rolf Dobelli. Mr. Dobelli, the driving force behind the popular business book summary website getAbstract (www.getabstract.com), is a writer in his own right. In an online essay titled "Avoid News," he recommends going cold turkey:

> *Make news as inaccessible as possible. Delete the news apps from your iPhone. Sell your TV. Cancel your newspaper subscriptions. Do not pick up newspapers and magazines that lie around in airports and train stations. Do not set your browser default to a news site. Pick a site that never changes. The more stale the better. Delete all news sites from your browser's favorites list. Delete the news widgets from your desktop.*[2]

As admirable as Mr. Dobelli's goal is, it is also unrealistic. Ever since Adam, humans succumb to temptation, and so are virtually incapable of going cold turkey. Why do you think there are so many diet books on the bestseller lists?

A more realistic approach to CrackBerry Addiction is to follow the model of other established substance abuse solutions: one step at a time. Mr. Perrottet tells of other authors who use a computer program called "Freedom" that blocks Internet access for up to eight hours.[3] Just as airlines require passengers to turn off their mobile devices for the duration of the flight, the withdrawal is confined to a limited period.

This is the first step toward self-control, and if self-control *ever* catches on, perhaps your next audience will take their eyes off their CrackBerries and focus on you.

Sure, and there *is* a Santa Claus.

55

The Eyes Have It

Relax!

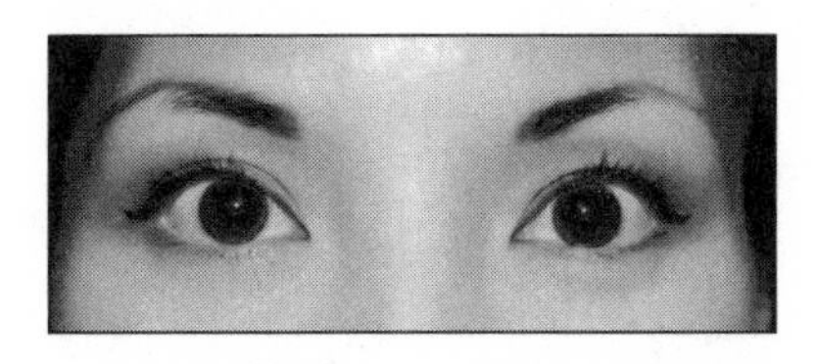

- "I *like* that person; he/she looks you straight in the eye!"
- "I *don't* like that person; he/she is shifty-eyed!"

These two familiar exclamations define the polar opposites of eye contact, the most essential element in interpersonal communication. But effective eye contact has another little-known but important benefit: calming the user.

Whenever you—or any presenter—stand up in front of any audience, the stress of the moment triggers an adrenaline rush that sets your whole body into the accelerated motion of Fight or Flight, particularly your eyes, which sweep the room in search of escape routes.

The rapid eye movement makes you appear furtive to your audience, which makes feel them uneasy; when you sense their uneasiness, you become more stressed, which heightens your adrenaline rush, which makes your eyes move faster and...a vicious cycle.

It gets worse.

During the sweep, your eyes take in a great deal of sensory data. All of that data must be processed by your brain, which increases your stress, which heightens your adrenaline rush, which makes your eyes sweep faster; the faster your eyes move, the more data you take in... the vicious cycle compounds.

Instead, look at each individual in your audience long enough to see that person look back at you. This simple step diminishes your rapid eye movement. Readers of *The Power Presenter* will recognize this technique as "Eye Connect," a more pronounced form of eye contact. With Eye Connect, you engage fully with each member of your audience. Contrast this approach with the scanning that most presenters do in their attempt to make eye contact. Connect with every person you see by waiting until you see that person look back at you, by waiting until you make the connection.

While Eye Connect decreases the frequency of your eye movement, it also decreases the amount of sensory data your brain has to process, which reduces your stress, lowers your adrenaline rush, and makes you calmer.

The calming effect created by diminished eye movement has an analogy in scuba diving. Karyn Scott, the Director of Enterprise Segment Marketing at Cisco, is a certified scuba diver. She explains that when she sees a novice diver panic under water, she swims to that person and gives hand signals—pointing two fingers rapidly back and forth between that person's eyes and hers—directing the person to look her in the eye. As soon as the person's eyes stop darting, his or her panic subsides, and the air bubbles coming from the diver's regulator quickly slow down. Connecting eye to eye with another human is so powerful there's almost no need for words.

Bruce Iliff, an Australian scuba Divemaster, has a variation of Ms. Scott's method: He recommends that when divers start to panic, they should "look at the surface. At 20 metres the surface looks so close you could reach out and touch it, a comforting thought!"[1]

In essence, both Mr. Iliff and Ms. Scott are advocating the same method you can use when you present: Look at each person in your audience until you see that person look back. That simple but powerful step will decrease the frequency of your eye movement, increase the duration of your engagement—and make you calmer.

That's Eye Connect.

56

Why Sinatra Stood

The Voice of "The Voice"

The legendary crooner Frank Sinatra had several nicknames, chief among them was "The Voice"; and for good reason: His rich, resonant vocal quality was without equal. "The Voice" took very good care to optimize his golden sound. One of Mr. Sinatra's standard techniques—practiced by *all* professional singers, as well as announcers, actors, narrators, and other performers whose voices are their livelihood—was to stand when he recorded his songs.

Look at any photograph of any singer in a recording studio, and you will see that they are standing upright in a soundproof booth with a microphone suspended in front them. Their sheet music is propped on a stand raised to the singer's eye level. This arrangement provides a valuable lesson for businesspeople who present over the web or on a telephone, and particularly on a speakerphone.

When anyone—including any presenter, our primary focus—talks into the handset of a telephone or a speakerphone while seated, that person will naturally lean forward, hunching the shoulders, constricting the lungs, and reducing the air supply. Leaning forward also drops the chin and constricts the throat, reducing the air supply even further. Standing up straight expands the lungs to fill with more air and lifts the chin to open the throat, creating an open channel for a full column of air.

Some businesspeople, who present online frequently, take this matter into account. Will Flash of Microsoft Corporation, who now manages a group of internal studios for the company, was previously the Broadcast Host of Microsoft's web meeting platform (formerly LiveMeeting, now Lync[1]). That role as called for him to present online several times a day. Each time he did, Mr. Flash stood up and pushed a button to elevate his motorized desk. In an instant, he replicated the configuration of a professional sound recording studio: By standing, he expanded his lungs and by looking at his computer screen at eye level, he opened his throat.

You don't have build a soundproof studio or install a motorized desk in your office, but in this globalized world where online presentations have become standard operating procedure for many businesses, be sure to take these simple steps to optimize your voice. Absent the essential elements of eye contact, facial expressions, and gestures that you use to augment your story during in-person presentations, your voice becomes the primary vehicle of delivery when you present on the web.

Make *your* story sing!

57

Presentation Counts

The Rise and Fall of Rick Perry

The first ever televised debate between U.S. presidential candidates Senator John F. Kennedy and incumbent Vice-President Richard M. Nixon in 1960 was the seminal event that changed the face of political campaigning forever. Mr. Kennedy entered the race as an underdog with two strikes against him: At 43, he was the second youngest man ever to run for office (William Jennings Bryant was 36 when he ran in 1896, and he lost to William McKinley) and the second Catholic candidate (the first, New York Governor Al Smith, had lost to Herbert Hoover 32 years earlier).

In the debate, Mr. Kennedy outshone Mr. Nixon and went on to win the election. Ever since, no candidate could succeed without having or acquiring powerful delivery skills.

- In the 1980s, The Great Communicator, Ronald Reagan, outshone both the homespun Jimmy Carter and the bland Walter Mondale in consecutive elections.
- In the 1990s, the glib Bill Clinton outshone both the patrician George H. W. Bush and the drab Bob Dole in consecutive elections.
- In 2008, the dynamic Barack Obama outshone the cantankerous John McCain.

Presentation counts.

The race for the Republican candidate for the 2012 presidential election began early in 2011. In May of that year, a crowded field of candidates engaged in the first of what would ultimately amount to 27 debates on national television.[1]

When Texas Governor Rick Perry entered the race on August 13, he was late to the game. Former Governor Mitt Romney, ultimately the winning candidate, led a large pack that included Former House Speaker Newt Gingrich, Former Senator Rick Santorum, Representative Ron Paul, Representative Michele Bachmann, businessman Herman Cain, and Former Governor Jon Huntsman.

But Mr. Perry brought a lot to the game. He was movie-star handsome with a square jaw, wavy hair, and an athlete's build clad in impeccably-tailored suits. He had won three consecutive gubernatorial races and was the darling of the influential conservative wing of the Republican Party that financed his campaign with an abundant war chest. Brandishing this powerful array of personal and political assets, Mr. Perry jumped to the head of the pack in the public opinion polls, only one month after his announcement, and joined the debates.

In the next debate on Fox News on September 22, correspondent Chris Wallace, true to his role as a moderator (and his legacy as the son of the renowned journalistic provocateur Mike Wallace) tried to stir up conflict. He asked Mr. Perry to comment on Mr. Romney's current health care proposal and the one he developed when he was the governor of Massachusetts:

> *PERRY: I think Americans just don't know sometimes which Mitt Romney they're dealing with. Is it the Mitt Romney that was on the side of against the Second Amendment before he was for the Second Amendment?*
>
> *Was it—was before he was before the social programs, from the standpoint of he was for standing up for Roe v. Wade before he was against Roe v. Wade? He was for Race to the Top, he's for Obamacare, and now he's against it. I mean, we'll wait until tomorrow and—and see which Mitt Romney we're really talking to tonight.*[2]

The convoluted ramble went viral in the media and, within two weeks, by October 8, Mr. Perry's numbers in the polls plummeted to almost half. You can see the results in Figure 57.1, taken from figures on the realclearpolitics.com website of their "RCP Poll Average for the 2012 Republican Presidential Nomination." Please note that we've charted Mr. Perry's progress only against that of Mr. Romney; for the purposes of this chapter, we have omitted the other candidates.[3]

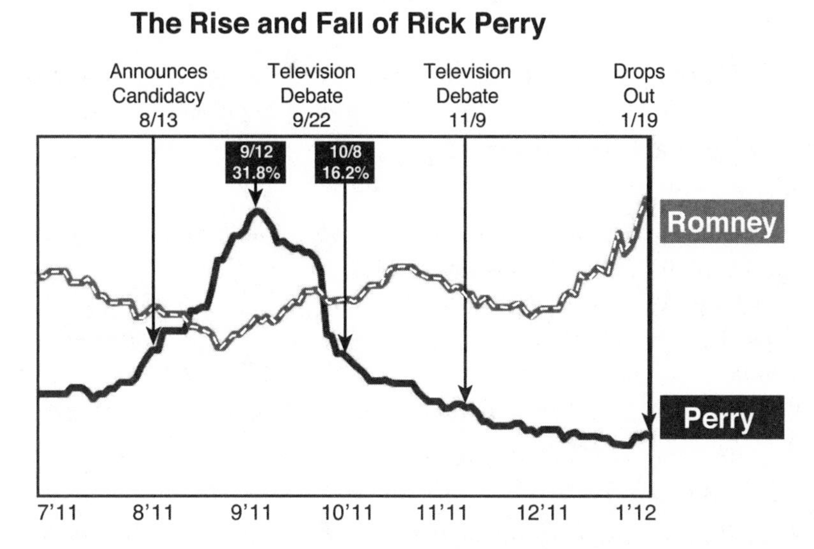

Figure 57.1 RCP Poll Average for the 2012 Republican Presidential Nomination

Mr. Perry participated in two more debates in October without incident. Then on November 9, he joined his opponents for another debate on CNBC, this one focused on the economy. During one of his turns, Mr. Perry explained his tax plan:

> *PERRY: But the fact of the matter is, we better have a plan in place that Americans can get their hands around, and that's the reason my flat tax is the only one of all the folks—these good folks on the stage. It balances the budget in 2020. It does the things for the regulatory climate that has to happen.*
>
> *And I will tell you, it's three agencies of government, when I get there, that are gone: Commerce, Education, and the—what's the third one there—let's see...*

Mr. Perry paused, unable to name the third agency he would eliminate. The audience began laughing. Ron Paul, who was standing next to Mr. Perry, made light of the moment, "You need five."

Mr. Perry tried to be light too, "Oh, five. OK," he said. Then, trying to remember the third agency, he started back at the beginning of his list, "OK. So Commerce, Education and...the...." But he paused again. The audience continued laughing, now joined by the other candidates.

Mitt Romney, also trying to be light, said with mock helpfulness, "EPA?"

Trying to be good-natured, Mr. Perry said, "EPA. There you go," waving his hand at Mr. Romney and then slapping it on the lectern in mock agreement. The gestures brought more laughter and a round of applause from the audience.

John Harwood, one of the CNBC moderators asked, "Seriously? Is EPA the one you were talking about?"

PERRY: No, sir. No, sir. We were talking about the agencies of government—EPA needs to be rebuilt. There's no doubt about that.

HARWOOD: But you can't—but you can't name the third one?

PERRY: The third agency of government.

HARWOOD: Yes.

PERRY: I would do away with the Education, the Commerce and—let's see—I can't.

He paused again, crestfallen.

PERRY: The third one, I can't. Sorry. Oops.[4]

The "Oops" heard around the world, along with the convoluted ramble of the September debate, proved to be Mr. Perry's undoing. His rise and fall forms an almost a perfect bell curve in the opinion polls, with the down side paralleling the two fateful gaffes in the debates. The slide continued until Mr. Perry withdrew his candidacy in January 2012.

Granted that the stakes in presidential debates are an order of magnitude greater than any business presentation you will ever give. But there is a significant parallel, and it was eloquently stated by Jon Meacham, a Pulitzer Prize-winning author who has written biographies of Andrew Jackson, Franklin Roosevelt, and Winston Churchill:

What is deeply relevant and often revealed in the debate formats that have become so familiar to Americans in presidential politics since 1960 is performance skill—the capacity to project a vision or communicate the essence of one's character with brief moments and gestures that create lasting impressions.[5]

Mr. Meacham's words are as deeply relevant to business communications as they are presidential politics: presentation counts.

Section IV

How to Handle Tough Questions

58

Listening and Laughing with Johnny Carson

Late Night Lessons for Presenters

In *Presentations in Action*, you read how Johnny Carson, the legendary late night television star, serves as a role model for two critical factors in presentations: listening and the danger of humor. Those lessons were reinforced by the release of a compilation of selections of Mr. Carson's television programs called, *Tonight: 4 Decades of The Tonight Show starring Johnny Carson*.

The collection was reviewed by the television critic for the *San Francisco Chronicle*, David Wiegand, who described Mr. Carson's listening skills perfectly:

> *It wasn't so much that Carson was a good interviewer—it was that he was a superb listener, giddier when one of his guests got off a good line than he would have been if he'd said the line himself.*
>
> *What difference does it make if a late-night host is a good listener or not? More than might be apparent, especially today when it seems as though the rule book for late-night hosts dictates that you display your own cleverness and, too often, snarkiness, at the expense of everything else.*
>
> *By contrast, a host who's a good listener becomes a real stand-in for the home audience. By showing himself more interested in the guests than finding a way to get in a good line himself, Carson made us feel as if we were sitting next to him at the desk asking the questions and, more important, interested in knowing the answer.*[1]

Mr. Wiegand's observations were seconded by his colleague, Dorothy Rabinowitz, the television critic of the *Wall Street Journal,* who summarized Mr. Carson's listening skills in one sentence:

> *This was not a show host in a hurry to get himself back to stage center.*[2]

Effective listening strikes at the essence of two aspects of presentations:

- Listen as you are asked questions. Make sure that you identify what your audience wants to know *before* you answer.
- Read the reaction of your audience for non-verbal signals—consider it silent listening. See whether your answer is producing head nods. If not, amplify your answer.

In both cases, your goal is to make your audience feel what Mr. Carson made his audiences feel, that you are more interested in your audience than in yourself.

Mr. Wiegand's observations about Mr. Carson's humor are also worth noting:

> *What's fascinating about a Carson monologue is how often the jokes fail to ignite, but, truth to tell, Carson was at his comedic best when the joke didn't work. To adapt a Mae West line, when he was good, he was funny; when he was bad, he was better.*

I hope this puts to rest that pervasive notion that one should start a presentation with a joke. If Mr. Carson, whose jokes were written by a crack team of professional comedy writers, bombed occasionally, why should you even try?

I also hope that you'll bear with my repeating these lessons about listening and laughter. After Death by PowerPoint, the lack of listening and failed attempts at humor rank high on the list of common mistakes presenters make—repeatedly. You can avoid these mistakes by learning from the King of Late Night Television. Johnny Carson won his rating period every year for 30 consecutive years.[3]

When you present, present to win.

59

Ready, Fire, Aim!

Old Habits Die Hard

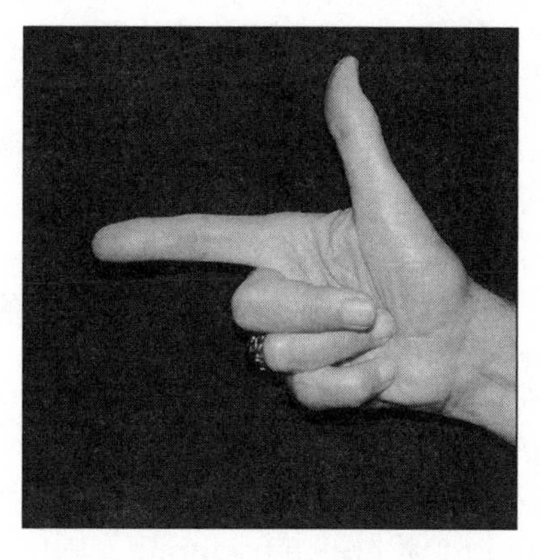

The quality that earned you your present job—as well as all of your previous jobs—is the same quality that impedes your ability to answer questions effectively: You are a results-driven person.

To determine how good and—just as important—how fast you are at producing results, your employer undoubtedly assessed your resume, your references, and your character during your intake interview.

Having demonstrated your proficiency means that you spend most of your time on your job (and most likely, the rest of your waking hours) ready to pounce on problems and find solutions. As a result, whenever you get a question, you are primed to provide an answer instantly.

Unfortunately, if you are too quick with your response, your answer might be wrong—because you did not understand the question. You will have fallen into the "Ready, Fire, Aim!" trap.

Being a good listener, as you learned from the example of Johnny Carson in the previous chapter, is important, but that is only the first step; it is just as important to take a beat before pulling the answer trigger; to put the aiming in its rightful place—*before* the firing. Pause before you answer a question.

But pausing is difficult during a mission-critical presentation because you are acting under the influence of a double speed whammy: the adrenaline overdrive of being on the spot and the DNA of a trigger-happy problem-solver.

Apple Computer understands trigger-happiness. The company, which is well-known for carefully guarding its product development, makes a practice of keeping all but a few select senior executives from answering questions from the press. In *Inside Apple*, a book describing what the subtitle calls its "secretive" practices, author and *Fortune Magazine* Senior Editor Adam Lashinsky quotes an Apple product marketing executive: "The challenge with those guys is that they are super smart and they know a lot of details, but...they haven't learned how to gracefully avoid answering."[1]

Apple's competitor, Google, also understands trigger-happiness. Their Gmail has a feature called "Undo Send." Once you hit "Send," Gmail will hold your e-mail for 5 seconds, during which time you can stop the e-mail from going out.[2]

The most sensitive trigger-happy arena of all is television and radio. Broadcasters employ a 7-second delay in live programs to monitor and edit undesirable material. Think of the wardrobe-malfunction at the Super Bowl or an excited blurt of profanity during a live Academy Awards acceptance speech. The most common example of the 7-second delay is the frequent sound of beeps during broadcasts of Jon Stewart's bawdy "The Daily Show."

Author Frank Partnoy extols the benefits of delay in his book, *Wait: The Art and Science of Delay*, with examples of a baseball batter waiting for the perfect pitch to hit, a comic waiting a beat before delivering a punch line, and a matchmaker counseling a blind date to suppress snap judgments.[3]

Because presenters do not have the facility of an "Undo Send" button or a 7-second video delay, they must create their own beat in real time—with a verbal pause. That is not a contradiction in terms; a verbal pause is none other than the old reliable paraphrase. Readers of *In the Line of Fire* will recall that the paraphrase—a reconfiguration of a question—serves as a *buffer*. For our purposes, let's consider the paraphrase as the presentation equivalent of the 7-second video delay.

For instance, suppose an irate customer were to ask, "Where do you get off charging so much for your product?" The hair trigger answerer might say, "It's not that expensive when you think of all the features you get..." or "You have to consider the long-term cost of ownership." Each of those rapid responses accuses the questioner of being wrong. If instead you paraphrase by saying, "Why have we chosen this price point?" you remain neutral—and you take that vital verbal beat.

Having bought the time, you can then go on to describe the features and/or the long-term cost of ownership, but you will do so without—to extend the metaphor—the crossfire.

Some presenters, in their quest to buy time when they get challenging questions, have developed a set of preliminary phrases to their paraphrases. Functioning as *double buffers*, the most common are

- "That was a good question!"
- "I'm glad you asked that!"
- "What you're really asking..."
- "If I understand your question..."
- "The issue/concern is..."

Sound familiar? These boilerplate phrases, like hair trigger answers, are deeply ingrained old habits. Old habits die hard, but die they must because each of them—one more extension of the metaphor—backfires.

- The first two phrases pass approval on the value of the question and, by implication, disapproval of the value of all other questions.
- The word "really" implies that the questioner cannot state his or her own question clearly.
- "If I understand your question..." implies that you were not listening.
- The words "issue" and "concern" bring forward a negative balance.

If you want an extra beat before your paraphrase, try these neutral and audience-friendly double buffers:

- "You're asking..."
- "You'd like to know..."
- "Your question is..."

Hair trigger answers and boilerplate buffers are old habits, as unproductive a habit as procrastination; replace them as you would any bad habit, with positive action: listening and paraphrasing—and get positive results.

60

How to Deal with a Direct Attack

"That was certainly a downer!"

The annual shareholder meeting of Yahoo! Inc. in June, 2011, was an "unlikely lovefest"—according to the report of the event in the *New York Times*[1]—unlikely because the company's stagnant stock price and poor performance against its competitors would seem to have invited a more critical session. The lovefest carried through more than an hour of glowing outlook from the board and management as well as softball questions from the shareholders.

But then a man named Steve Landry took the floor and identified himself as a personal investor who also advised institutional investors holding more than a million shares of Yahoo! Mr. Landry then proceeded to attack the company in general and its Chief Executive Officer, Carol Bartz, in particular. Addressing her directly, he said, "It came out earlier this week in a blog that the board is secretly talking to other potential CEO candidates. I've heard similar details and believe that it's true." He then went on to say, "the last thing Yahoo! needs is a lame duck CEO.... The buyout talks of your contract need to start today and a search needs to be accelerated."

According to the transcript of the webcast, Ms. Bartz responded:

> *Thanks for your opinion, the bloggers and the rumors. What else? Wonderful—that was certainly a downer. So again, thank you for coming to Yahoo! We are working very, very hard in this company and managing our assets and we will see the benefit of that.*[2]

However, the reports of the exchange in the major financial press—the *Wall Street Journal*,[3] the *New York Times*,[4] and

CNET[5]—carried only the central phrase of Ms. Bartz' answer, "that was certainly a downer." This left readers with the impression that Mr. Landry's direct attack had damaged her.

While Ms. Bartz did the right thing in ending her answer on an upbeat note, she made two tactical errors at the front end of her response, both of which outweighed and therefore overshadowed her positive effort.

- She neglected to address the issue in the attack.
- She validated the negativity.

Instead, Ms. Bartz should have addressed the issue of the rumors by saying:

Yahoo!'s policy is not to comment on rumors and blogs. What I can tell you is that...

And then she could have gone on to conclude with her original upbeat ending, "We are working very, very hard in this company and managing our assets and we will see the benefit of that."

These two issues provide larger lessons for how any presenter should handle tough questions or direct attacks.

1. Presenters have every right to take every opportunity to make positive statements about their companies, but they must first earn that right by addressing the central issue in the challenger's tough question or statement. They may agree, disagree, admit, correct, or deny the issue, but presenters cannot ignore it. Ms. Bartz would have addressed the issue of rumors by stating Yahoo!'s policy on rumors.
2. Negative facts may be sad but true—Yahoo!'s stock price had been stagnant—and presenters can actually be forthright about it, but they should not validate the negativity. "That was certainly a downer" carried forward the negative balance.

   Instead, presenters can admit to negative facts—so as to be accountable—but then, after a brief, *very* brief, admission, immediately follow it with an upbeat counterpunch. This known as the "Yes, but..." approach:

   Yes, the stock price has been stagnant, but when you consider the outlook for the new products that you've heard about today and the fact that we are working very, very hard in this company and managing our assets, I am confident that we will see the benefit of that.

Following the annual meeting, CNET reported that a Yahoo! spokesperson emailed to say, "Rumors suggesting there is or has been any sort of search for a replacement to Carol are categorically untrue."[6]

Three months after that, however, with the stock price still stagnant, the Board of Directors did fire Ms. Bartz. Undoubtedly, it was the performance of the stock and not her testy performance at the annual meeting that brought about her dismissal. As the *New York Times* report of the firing noted, "She has long been inclined to honesty, often in salty language."[7]

Carol Bartz can afford to be direct. Before Yahoo!, she was the CEO of Autodesk for 14 years and has served on the Boards of Directors of Intel, Cisco, BEA Systems, and Network Appliance. Unless you have that kind of track record, when you are asked a tough question, you must:

- Address the issue
- Avoid negativity
- Be accountable
- End upbeat

61

No Such Thing as a Stupid Question

A Lesson in Q&A from Dilbert

The *Dilbert* comic strip which lampoons all aspects of the business world, turned its attention to Q&A with a panel that had Dogbert, one of the strip's recurring characters, sitting under a red banner reading "Communication Skills Training." Dogbert is addressing a class and says, "Today you will learn how to listen to idiots without snorting."[1]

Dogbert was referring to the spontaneous reaction that many people in business display when they are asked what they consider to be idiotic or stupid questions. But this disdainful behavior can boomerang because the person who asked the question does not consider it stupid. And that person could very well be the decision maker who can either approve or disapprove of a presenter's pitch.

Scott Adams, the creator of *Dilbert*, wielding his usual satiric blade, was cutting both ways: mocking seemingly stupid questions and showing that the snorting reaction is equally stupid.

In the business world, there is no such thing as a stupid question. A question might be uninformed, tangential, or seemingly irrelevant, but, whether the presenter perceives it to be stupid or not, every audience member has every right to ask any kind of question. To adapt the famous line from the 1989 film *Field of Dreams*, if they ask it, you must deal with it—but you must deal with it without offending.

When fielding questions from your audience, the correct way to react to a question that *you* might perceive to be off-topic is to immediately paraphrase the key issue. In Chapter 59, "Ready, Fire, Aim!", you read how the paraphrase neutralizes challenging questions from the audience—it can also neutralize disdainful responses from the presenter.

For instance, if at the end of a presentation about the bells and whistles of your new product, you were to be asked about the typeface of your logo, instead of yielding to the common temptation to say, "What does that have to do with the product?" say instead, "How did we choose our brand design?"

This simple step will not only inhibit any snorts—or snickers, smirks, frowns, nose-wrinkling, or eye-rolling—it will lead you to a respectful answer to your valued audience member.

62

The Patronizing Paraphrase

Trying to Channel Bill Clinton

Scenario #1: *Silicon Valley, an Executive Briefing Center at a major IT company.* One of the company's product managers finishes a presentation about a product upgrade to a group of existing customers and then opens the floor to questions.

The first question comes from the CIO of a large financial institution: "We've spent millions of dollars on the first version of your solution, and it gave us nothing but problems—crashes, down time, glitches, and endless repairs—and now you want us to upgrade to a new version! We're still having problems with the earlier version. What are you folks going to do about it?"

The product manager responds, "Quality is important to us...."

Scenario #2: *New York City, a hotel banquet room during a financial conference.* A CEO of a public company finishes the company's management presentation to a group of investors and then opens the floor to questions.

The first question comes from an analyst at a leading mutual fund: "Your revenues are flat, your stock is down, and your outlook for the next quarter is guarded. When are you going to turn this sucker around?"

The CEO responds, "Performance is important to us...."

Scenario #3: *Chicago, a conference room at the headquarters of a national retail chain.* An account executive of a manufacturing company finishes a presentation about the status of a current product and then opens the floor to questions.

The first question comes from the vice president of sales: "Your last product was late and the one before that was late. Now you tell us that this one will be late! You know that our sales are seasonal and if we miss that narrow window we lose revenues and market share. When are you guys going to get your act together?"

The account executive responds, "Promptness is important to us...."

Sound familiar? No doubt you've probably heard the "(Blank) is important to us" phrase countless times. It has become boilerplate in Q&A sessions.

The problem with the phrase is that it is the blinding flash of the obvious. Of course quality, performance, and promptness are important—each of the questioners just got finished saying that! Therefore, when a presenter states the obvious in a paraphrase, it is patronizing to the audience.

Why would any presenter do that to any audience? It is probably a misguided attempt to echo Bill Clinton's famous words, "I feel your pain." Mr. Clinton coined the phrase during his run for the presidency in 1992 in response to a question from an AIDS victim. The phrase was to become a campaign slogan that sent a broader message that Mr. Clinton hears and understands every voter.

As a presenter, it is vitally important that you send the message that you hear and understand every questioner, but do so without saying that you feel your audience's pain—especially when, by the challenging nature of the question, you or your company caused the pain in the first place.

Instead, use the paraphrase correctly: to neutralize the key issue—not to share feelings, not to agree, and certainly not to patronize.

The correct paraphrase for each of the three tough questions above is:

- "What we're doing to assure quality is...."
- "What we're doing to improve performance is...."
- "What we're doing about on time delivery is...."

From this neutral start, you can move forward into an answer as to how you are going to address the questioner's problem. And be sure that your answer addresses the problem thoroughly.

If you want to channel Bill Clinton's undoubtedly effective presentation style, follow the advice of his campaign slogan, "Put People First."

63

Tricky Questions

Be Transparent or Be Trapped

Adam Bryant interviews CEOs for his weekly "Corner Office" column in the *New York Times*. He distilled a group of the interviews into a bestselling book called *The Corner Office,* whose subtitle describes its intent: *Indispensable and Unexpected Lessons from CEOs on How to Lead and Succeed.*

The key word is "lessons." I follow Mr. Bryant's excellent column every Sunday and often find valuable lessons for presenters. Two of those columns addressed the same subject: how to handle tricky questions: Mr. Bryant asked two chief executives—Kevin O'Connor of FindTheBest.com and Paul Maritz of VMware—about the tough questions they ask prospective employees. Both men referenced questions that, on the surface, appeared to be deceptively easy.

Mr. O'Connor said, "I try to give them a question that feels like a two-by-four between the eyes," and he cited as an example, "How smart are you?"[1]

At first, this question may sound more like a free pass than a two-by-four, but it can become a trap for the latter. "How smart are you?" is analogous to one of the most frequently asked presentation questions, "What keeps you up at night?" If a presenter were to reply, "Nothing keeps me up. I sleep like a baby," the presenter would appear to be unrealistic or evasive. Audiences want presenters to be candid, but also to describe what they are doing about problems.

In that same way, if an interviewee were to reply to the "How smart are you?" question with, "I'm the best in the business!" that would make that person appear to be hyperbolic.

When Mr. Bryant asked Mr. O'Connor what he would consider a good response to this question, Mr. O'Connor replied, "someone who's really, really good at something, and knows it. But they also realize they have shortcomings in other areas."

And when Mr. Bryant asked Mr. Maritz about his tough interview questions, he replied, "I'll just pick anything that they've done in the past and I'll say: 'Thinking about it now, what would you have done differently? What did you learn from that?'"[2]

Just like Mr. O'Connor, Mr. Maritz looks for realistic or direct answers. He said,

> *If they blame everything that happened during that period on somebody else, that tells you that the person is probably not thoughtful or self-aware. If they can talk in length about what was really going on, why they made the decisions they did and how they would perhaps make the decision differently now, that tells you that this person thinks deeply and is honest enough to really be objective, or as objective as they can be about themselves.*

The common denominator that both Mr. O'Connor and Mr. Maritz seek is transparency expressed with honesty and candor. Interviewees and presenters (as well as all human beings) must be straightforward about their limitations. Once they acknowledge that these exist, they are free to go on to state their own case. Not the other way around. Politicians, whose perpetual spinning the public has come to tolerate, are perpetually stating their own cases and are rarely transparent—the political equivalent of "I'm the best in the business!" type hype.

This is not to say that you shouldn't toot your own horn; by all means do so, but do it with honesty and candor.

64

Robert McNamara Was Wrong

You Must *Respond to All Questions*

Robert S. McNamara was the Secretary of Defense from 1961 to 1968 and the driving force behind the controversial Vietnam War. He went on to a more successful stint as head of the World Bank and lived until the ripe old age of 93, but according to his *New York Times* obituary, "spent the rest of his life wrestling with the war's moral consequences."[1]

As part of his struggle, he agreed to be the subject of a 2003 documentary in which he expressed regrets but ultimately defended his actions. The film is called *The Fog of War: Eleven Lessons of Robert S. McNamara*. Lesson Ten is about communication, and it contains sound advice for presenters about what *not* to do. Mr. McNamara said:

> *One of the lessons I learned early on: never say never. Never, never, never. Never say never. And secondly, never answer the question that is asked of you. Answer the question that you wish had been asked of you. And quite frankly, I follow that rule. It's a very good rule.*[2]

It is a very bad rule, and unfortunately it has taken on a life of its own in the modern business world. Many media consultants urge presenters to answer "the question you wish had been asked," and to deliver their own message. Yes, it's good to do that—within bounds. But think about it: How can it be a "very good rule" not to be responsive to other people?

In interpersonal relationships, not answering a question can lead to an argument; in business, not answering a question can lead to the failure of a deal. Only in politics, where the public has become inured to the practice of ducking and spinning, do audiences tolerate unanswered questions. But even there, the McNamara rule can backfire.

In the contest for the 2012 Republican presidential nomination, as you read in Chapter 57, "Presentation Counts," Texas Governor Rick Perry had a rapid rise and fall due in large part to his notorious performance blunders in two debates. But what was overlooked in all that attention was a Robert McNamara moment in the October 18, 2011, debate on CNN, when moderator Anderson Cooper asked Mr. Perry this question:

COOPER: Governor Perry, the 14th Amendment allows anybody. A child of illegal immigrants who is born here is automatically an American citizen. Should that change?

PERRY: Well, let me address Herman's issue that he just talked about.

COOPER: Actually, I'd rather you answer that question.

PERRY: I understand that. You get to ask the questions, I get to answer like I want to.[3]

"I get to answer like I want to." Imagine a salesperson saying that to a customer, a manger to a senior executive, a senior executive to a board member, or a CEO to an investor. Meeting over. No deal.

Imagine saying that to your significant other. No comment.

Anderson Cooper called Mr. Perry on it, "That's actually a response; that's not an answer."

Four months later, in another debate among Republican candidates, another Robert McNamara moment occurred in this exchange between Mitt Romney and CNN moderator John King:

KING: What is the biggest misconception about you in the public debate right now?

ROMNEY: We've got to restore America's promise in this country where people know that with hard work and education, that they're going to be secure and prosperous and that their kids will have a brighter future than they've had. For that to happen, we're going to have to have dramatic fundamental change in Washington, D.C., we're going to have to create more jobs, have less debt, and shrink the size of the government. I'm the only person in this race—

KING: Is there a misconception about you? The question is a misconception.

ROMNEY: You know, you get to ask the questions you want; I get to give the answers I want.[4]

In business, you do not have the luxury of that kind of answer. You must respond to all questions. This is not to say that you should give away state secrets; you have every right to decline to answer on the basis of confidentiality, competitive data, or company or legal policy, but you must provide a rational reason—and "I get to answer like I want to" is irrational.

65

Breaking into Jail

The Elephant IS in the Room

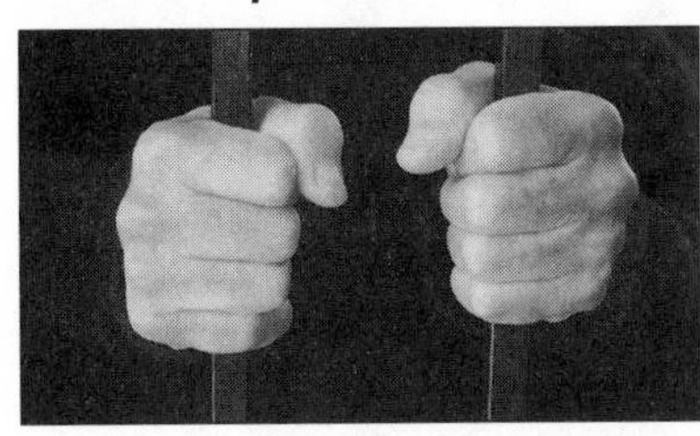

Jail break films have long been a staple offering of Hollywood, but in a real life reversal of form, the *Los Angeles Times* reported that a former prisoner of the California State Prison in Sacramento, one Marvin Ussery, attempted to break *into* the jail. Although Mr. Ussery claimed that he was only "reminiscing," prison authorities suspected that he was trying to smuggle in drugs, tobacco, or mobile phones to sell to the inmates. However, a search didn't find any contraband on him, so his motive remains a mystery.[1]

In business, "breaking into jail" has a different connotation: offering negative information voluntarily. Revealing a liability raises doubts in the audience's minds about a company's viability. However, there is a very good reason for such revelations: *accountability*. In presentations, unlike awkward social situations, the elephant in the room cannot and must not be ignored.

In some cases, accountability is mandatory. The Securities and Exchange Commission requires that a company selling stock to the public for the first time must include a "Risk Factors" section in the Prospectus for their Initial Public Offering. But the road show presentation of the offering isn't required to use the Draconian language of the prospectus. Nor is there such a requirement for countless other routine types of business presentations. Yet all presentations must be forthcoming about bad news, or the presenter will be perceived of as having something to hide.

The challenge is when and how to handle the revelation. The "when" has two options:

- **Be preemptive.** Include the negative information in the body of your presentation.
- **Be reactive.** Wait until a question comes from the audience and have a prepared response ready.

Each option has a risk. Offering negative information is "breaking into jail," or admitting guilt, and raises an issue that the audience may not have considered. Waiting until a question is asked can appear evasive or concealing.

Regardless of which option you choose—and the choice is a judgment call that depends on the situation, the audience, and/or the presenter—you must then make full disclosure by acknowledging the negative. But, as soon as you do, follow up immediately with the actions that you and your company are taking to rectify the problem or to prevent its recurrence.

If your bad news is about

- A down quarter, describe your extra efforts to stimulate new sales
- The loss of a key customer, explain your efforts to win a new customer
- The resignation of a key executive, talk about your search outreach
- A delayed product release, lay out your accelerated production schedule
- A failed product trial, list the corrections you are making
- A critical comment by an important thought leader, find a more positive opinion and quote that person.

This strategy is a variation of the correct method for handling the ritual "What keeps you up at night?" question. Be candid about what keeps you up at night, but immediately follow up with what you are doing about it. Be candid about your company's bad news, but immediately follow up with what you are doing about it.

Acknowledge that the elephant *is* in the room, and then lead it out.

Section V

Special Presentations

66

Speak Crisply and Eliminate Mumbling

Why can't the English learn to speak?

—"My Fair Lady,"
Lyrics by Alan Jay Lerner

Be Your Own Henry Higgins

How often have you had to ask someone to repeat what they just said? If you're like most people, probably quite often. Putting the shoe on the other foot, how often have *you* been asked to repeat what *you've* just said? If you're like most people, probably quite often.

This all-too-frequent social exchange is *not* due to a decline of the auditory capacity of our population; *the fault, Dear Brutus, is not in our stars, but in ourselves.* Mumbling has become chronic in conversation, in the theater, and in films. The *New York Times* described a new cinema genre called "Mumblecore," characterized by "low-key naturalism, low-fi production values and a stream of low-volume chatter often perceived as ineloquence."[1]

Such films, as well as many theater productions, cause you to turn to your companion to ask, "What'd he say?" In conversation, mumbling causes a similar request for repetition. In presentations and speeches, it causes your audience to miss or worse, misunderstand, your valuable words. Still worse, ineloquence creates a negative perception.

What's the rush? The culprit is the accelerated pace of our everyday lives. Powered by increasingly faster airplanes, vehicles, Internet speeds, devices, and apps, we are driven to perform all our mundane daily activities faster: traveling, eating, walking, tweeting—and speaking. This rushed rhythm has propelled the rate of our speech into unintelligible garbles.

Don't blame the twenty-first century. Consider the pre-technology nineteenth and twentieth centuries and the differing lifestyles—and speech cadences—of urban and rural dwellers. New Yorkers have always been characterized by their rapid speech, Southerners and Westerners by their slow drawl. Even in today's highly mobile society, those patterns continue to identify a person's origins. Many native New Yorkers still sound like auctioneers on steroids.

Unfortunately, the remedy is *not* to slow down. People cannot slow down; they can only control the tempo of their speech. Pausing helps to accomplish that goal, yet there is another way to facilitate control, a simple technique to eliminate mumbling and speak crisply.

Athletic Articulation

The great Irish-born British Nobel laureate George Bernard Shaw was a writer who knew a thing or two about voice, having created the character of Professor Henry Higgins, the quintessential speech teacher, for his play *Pygmalion*, the source for the world famous musical *My Fair Lady*. Mr. Shaw, a keen critic, was also highly attuned to how people spoke; he gave Professor Higgins eloquent dialogue.

In one of Mr. Shaw's earlier plays, *Candida,* he described one of the principal characters, The Reverend James Mavor Morell, as follows:

> *A vigorous, genial, popular man of forty, robust and good looking, full of energy, with pleasant, hearty, considerate manners, and a sound, unaffected voice, which he uses with the clean,* ***athletic articulation*** *of a practised orator, and with a wide range and perfect command of expression.*[2]

Note the key words "athletic articulation." The common definitions of those words are obvious; athletic refers to sports, and articulation to speaking. But each word has an additional definition: athletic means vigorous, and articulation means movement. Mr. Shaw's two simple words provide the remedy for mumbling: to speak as Reverend Morell does, with a "perfect command of expression," you must move your mouth, the primary mechanism of speech, vigorously.

Pardon the play on words, but vigorous movement is much easier said than done. If you mumble, you do so because, in your accelerated speech pattern, you move your mouth, lips, tongue, and jaw quickly in a tightly confined range of motion. This constricted action clips the production of individual words and jams the words against one another, creating that indistinguishable garble. Furthermore, having spoken rapidly all your life, you have developed a habit pattern as deeply ingrained as being right- or left-handed. Breaking that speed pattern is almost as difficult as switching handedness. It can be done, but you have to do it *progressively*.

Primer

To implement athletic articulation, hold up a mirror and watch your mouth as you speak. Notice the action of your lips and your jaw. Are they moving vigorously? Are you seeing your teeth as you speak? If you mumble, you will see very little white.

Now say the words, "athletic articulation," and notice what happens. You see more white. The words cause you to *do* what they *mean*.

Now say "athletic articulation" again, but this time, exaggerate each word and observe your teeth. More white. You'll also notice that your pronunciation of the words is clearer.

At this point you're probably thinking that all this exaggerated movement looks strange. That's right it does, and far be it for me to recommend that you act strangely to sound clear. But consider the exaggeration as a warm-up process, just like stretching before engaging in a sport. Athletes warm up differently from the way they perform in a game. Extending the analogy, do not attempt to perform this technique in your game of life, in social or business settings. Practice the exercise in private until you gain facility.

A simple rule of thumb is to repeat the words "athletic articulation" in front of a mirror twice a day for about a minute each time. You can do these exercises right after you brush your teeth in the morning and at night. (The white will be whiter.) After a while, begin to add some other random words either spontaneously or reading a sentence or two from a newspaper or magazine. Always focus on *seeing* and *feeling* the vigorous movement of your mouth. Also listen and you'll hear your words become crisper.

Seek your own comfort level, the point at which your movements feel less exaggerated, but your words sound more distinct. The time it takes to reach that level will vary by person and by the time and diligence each person applies.

When you feel comfortable, try to speak with athletic articulation in select private situations: with a family member, a friend or a co-worker. Little by little, as you gain more facility, you can begin to try athletic articulation in more open social settings. Do it in small increments. Walk before you run. Wait until you have attained mastery to try your new skill in an important presentation or speech. The operative word is *progressively*.

I practice what I preach. As a native New Yorker, I spoke my words at Olympic track speeds. Now I teach other people how to control their cadence.

Be Your Own Henry Higgins

One of the most memorable scenes in *My Fair Lady* is when Professor Higgins has a breakthrough in his efforts to train Eliza Doolittle, a Cockney flower girl, to speak as if she were a duchess. To build tension for the breakthrough, the staging of the scene makes it clear that the process has been long and arduous. As Eliza struggles to pronounce her exercise, "The rain in Spain stays mainly in the plain," an exhausted Professor Higgins slouches in his chair and repeatedly says, "Again."[3]

Your athletic articulation exercises, while tedious, do not need the services of a Professor Higgins. Eliza Doolittle was trying to alter her Cockney pronunciation, the *sound* of her voice, and she needed help.

We don't *hear* ourselves objectively. Use the practice recommendation in Chapter 13, "Writer's Block II." Develop your story by recording yourself and then listening to the recording to hear your speech pattern objectively.

In your practice to control the rate of your speech, you have two far more objective criteria than sound: the *sight* and the *feel* of your mouth, lips, and jaw in motion. If you practice in front of a mirror and *see* and *feel* your facial muscles and mouth moving vigorously—and your teeth flashing white—you will know that you are speaking crisply. At that point, you become your own Professor Higgins and can provide your own "Agains." Repetition over time is the only way you can effect real change.

You needn't aspire to become "a practised orator" like Reverend Morell or try to pass yourself off as a duchess as Eliza Doolittle did. Just be sure that every word you utter in your presentation or speech is crystal clear.

67

How to Develop a Richer Voice

Be Your Own Echo Chamber

Ever since Adam and Eve, the difference between men and women has been discussed and debated ad infinitum. In voice, however, the difference is indisputably clear: The larynx, or vocal cords, the principal component of the human voice, is smaller in females than in males; and so, just as a guitar's shorter and thinner strings produce a higher-pitched musical tone, a woman's shorter and thinner vocal cords—as well as those of a small man—produce a lighter, higher-pitched sound.

And so do the vocal cords of a teenager. *There's the rub*: The squeaky quality of an adolescent's voice is a mark of their immaturity, and so women and small men with thin, high-pitched voices are unfairly saddled with the same perception—a career-limiting handicap.

In an early scene in *The Iron Lady*, a biopic of former British Prime Minister Margaret Thatcher, Meryl Streep, who portrays Lady Thatcher, is told by her political consultants that her screechy voice lacks gravitas. The next scene is a montage of Ms. Streep/Thatcher being coached by a voice teacher who tries to get her to drop her tone by raising and lowering her extended arms.

This method is effective for pumping air from the lungs and adding volume to the voice, but it does little to the tone of the voice. The basis of high-pitched sound is not the air flow coming from the lungs, but from the vocal cords, which are situated *above* the lungs, and so the correct way to enrich the voice is to work from the larynx up by creating resonance.

Resonance

The word "resonance" combines the prefix "re-," meaning "repeat," and "sonance," meaning "sound." Repeat sound. An echo is a repetition of sound. Think of the times you've stood deep in a cave or high on a mountain overlooking a valley and said "Hello!" and then heard your own "Hello!" bounce back or *resound* in a deep boom.

The echo of your voice contained overtones and undertones that made your original "Hello!" sound deeper and richer. If you could add overtones and undertones *whenever* you speak you would develop a deeper and richer version of your own voice. You can create an echo with the mechanism of your own body without having to enter a cave, stand on a mountaintop, or haul around an amplification system.

Let's analyze the elements of the mountain echo: Your voice originated in your larynx, and then traveled across the open space of the cave or the valley. The sound then struck the walls of the cave or the sides of the valley and bounced back with renewed energy. The bounces are resonance, the reverberations that add the overtones and undertones that enrich the original sound.

The elements of that process were:

- Your voice
- The open space of the cave or the valley
- The walls of the cave or the sides of the valley

Or put another way:

- A source
- An open chamber
- A hard surface

You can incorporate those elements into your normal, everyday speaking voice because:

- Your larynx is the source of your voice.
- You have open chambers in your own body.
- You have hard surfaces in your own body.

Please note my restatement that the larynx is the *source* of your voice. Many schools of thought about vocal production focus on the role of the chest, the lungs, and the diaphragm—and for the voice coach in *The Iron Lady*, the arms—but they are *not* the source of voice. All those components do is generate a *silent* column of air that only becomes voice when it passes through the vocal cords. Sound does not occur until there is vibration.

Granted your chest, lungs, and diaphragm are open chambers with hard surfaces, but they are positioned *below* your larynx, and so there is no sound to resonate. If you look *above* the larynx, in Figure 67.1, you will see three open chambers or cavities, each of them is lined with hard tissue that provides a surface to bounce the vibrations that your larynx produces, to resonate the sound of your voice.

- Your throat, known as the pharynx, or the pharyngeal cavity
- Your mouth, the oral cavity
- Your sinuses, the nasal cavity

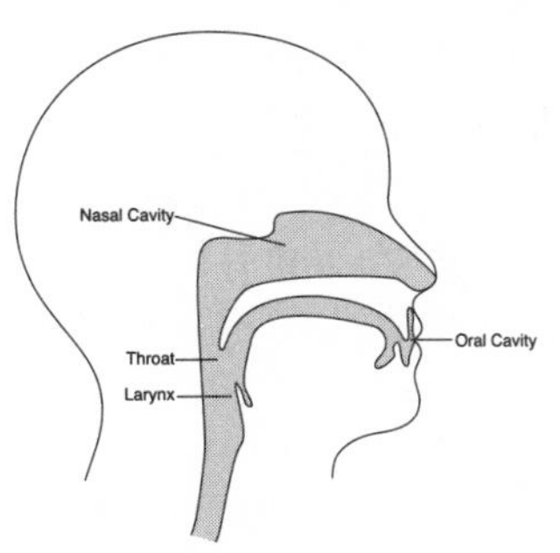

Figure 67.1 The pharyngeal, oral, and nasal cavities

As you speak, the vibrating air column of your voice passes through these three chambers anyway, but resonance is not a given. To create resonance you have to optimize the bounces by opening the chambers wider. The larger the chamber—the deeper the cave, the steeper the valley—the better the bounce, the greater the boom.

Be Your Own Echo Chamber

The vibrations produced by the larynx are expressed in two types of sounds: vowels and consonants. Vowels are usually identified as "a," "e," "i," "o," and "u," and consonants as every other letter in the alphabet. Actually, there are more than a dozen and a half different vowel *sounds* and countless consonant *sounds*. Rather than detail each one, however, let me simplify the definitions.

Consonants are sounds that are impeded by the actions of the lips, teeth, and tongue, while all vowel sounds are unimpeded or open sounds that are varied only by the shape of the mouth and the position of the jaw. Every vowel sound is the equivalent of your "Hello" on the mountain top, ready to reverberate into overtones and undertones.

Vowel sounds occur in every word we speak, even in Slavic languages, where some words do not have a single letter representing a vowel, but every word has a vowel *sound*. Therefore, you can use every vowel sound to add resonance to your every word.

Since every vowel sound travels through your throat and your mouth, these cavities can become your echo chambers for resonance. To incorporate the other available chamber, the nasal cavity, we turn to three other sounds: "m," "n," and "ng." Although these three are consonants—because they are impeded—they occur so frequently in our language, they provide you with an additional echo chamber and an additional opportunity to add resonance to your voice.

To develop nasal resonance:

Hum "mmm." Consciously focus on pushing the sound into your nose. Prolong the sound. Feel the vibrations in your nose and across your cheekbones.

Hum "nnn." Consciously focus on pushing the sound into your nose. Prolong the sound. Feel the vibrations in your nose and across your cheekbones.

Hum "ng," as in the end of "going." Consciously focus on pushing the sound into your nose. Prolong the sound. Feel the vibrations in your nose and across your cheekbones.

You have just experienced nasal resonance in isolation. Now say "many men marching." Concentrate on exaggerating the "m," "n," and "ng" sounds in each word.

Now say "moon beams bring evening dreams." Concentrate on exaggerating the "m," "n," and "ng" sounds.

> Now say "a strong new nation is going in motion." Concentrate on exaggerating the "m," "n," and "ng" sounds.

You have just experienced nasal resonance in controlled word strings.

> Now pick up a newspaper or magazine and read a paragraph aloud. Concentrate on exaggerating each "m," "n," and "ng" sound as it occurs in random words.

Practice these exercises daily until you find your own comfort level, the point at which your movements feel less exaggerated, but your voice sounds richer. The time it takes to reach that level will vary by person and by the time and diligence each person applies.

When you feel comfortable, try to speak with resonance in select private situations: with a family member, a friend, or a co-worker. Little by little, as you gain more facility, you can begin to try to resonate in more open social settings. Do it in small increments. Walk before you run. Wait until you have attained mastery to try your new skill in a mission-critical presentation or speech. The operative word is *progressively*.

To summarize, you now have a five-step process to develop nasal resonance:

1. Isolated exercises
2. Controlled word strings
3. Random exercises
4. Conversation
5. Presentation or speech

Steps 4 and 5 depend on your own rate of progress. Now let's apply the first three steps to oral and pharyngeal resonance.

To develop oral resonance:

> Say "ow," as in "how." Consciously focus on rounding your lips. Prolong the sound. Feel the reverberations in the front your mouth.
>
> Say "ah," as in "as." Consciously focus on dropping your jaw and placing the sound in the front of your mouth. Prolong the sound. Feel the reverberations in the front of your mouth.

You have just experienced oral resonance in isolation.

Now say "how now brown cow." Concentrate on exaggerating each vowel sound and feeling it reverberate in the front of your mouth.

Now say "so flow can grow slowly." Concentrate on exaggerating each vowel sound and feeling it reverberate in the front of your mouth.

Now say "as an afternoon advances." Concentrate on exaggerating each vowel sound and feeling it reverberate in the front of your mouth.

You have just experienced oral resonance in controlled word strings.

Now pick up a newspaper or magazine and read a paragraph aloud. Concentrate on exaggerating each "ow" and "ah" sound as it occurs in random words.

Now move on progressively to steps 4 and 5.

To develop pharyngeal resonance:

Say "on." Consciously focus on opening your throat and dropping your jaw. Prolong the sound. Feel the reverberations in your throat.

Say "oh." Consciously focus on opening your throat and dropping your jaw. Prolong the sound. Feel the reverberations in your throat.

You have just experienced pharyngeal resonance in isolation.

Now say "on an old island." Concentrate on exaggerating each vowel sound and feeling it reverberate in your throat.

Now say "a hot shot on the top." Concentrate on exaggerating each vowel sound and feeling it reverberate in your throat.

Now say "he ought to be caught and taught." Concentrate on exaggerating each vowel sound and feeling it reverberate in your throat.

You have just experienced pharyngeal resonance in controlled word strings.

Now pick up a newspaper or magazine and read a paragraph aloud. Concentrate on exaggerating each "on" and "oh" sound as it occurs in random words.

Now move on progressively to steps 4 and 5.

Your goal is to get the point where none of these techniques is a conscious effort; the point at which every word you speak gets channeled into its own chamber where it caroms around into overtones and undertones.

The result will be a rich, resonant voice that commands attention and gives *you* the power advantage.

The Pause Bonus

In Chapters 47 and 53, respectively, "Sounds of Silence" and "Foreign Films," you read about how pausing at the ends of your phrases gives your audience time to absorb your words. Pausing can also help to improve the quality of your voice.

If you speak rapidly, in a steady flat line string of words with few pauses, your vocal cords will vibrate rapidly and gain momentum. Rapid vibrations produce a high frequency or high-pitched sound.

If instead you break your word strings into logical bites—phrases—and punctuate them with pauses, you bring your vocal cords to a full stop at the end of each phrase. When you restart your next phrase, your vocal cords will start vibrating from a still position and take a short while to gain momentum. During this interval, your vocal cords will vibrate more slowly. Lower frequency produces a lower pitch.

Obey the traffic rules of good speech: Come to a full stop at the end of your phrases and pause.

Combine the lower pitch created by the pause and the overtones and undertones created by pharyngeal, oral, and nasal resonance, and you will speak with the voice of authority.

68

How to Deliver a Scripted Speech

When the Words Count

Microsoft PowerPoint, the graphic medium of choice for today's presentations and speeches, has also come to serve as a "script" or a prompt for the speaker's narrative. This convention gives the speaker the liberty to depart from the slides by either going deep or touching lightly on the main points as needed. However, there are several situations where the words in a speech must be more specific, where the "script" loses its quotation marks and becomes a prescribed text.

- **Legal.** A speaker having to make a public statement about a legal matter will use words carefully crafted by attorneys.
- **Policy.** A speaker having to make a statement about a controversial or sensitive issue will use words carefully positioned by public relations or media counsel.
- **Lyrical.** A speaker having to talk in an emotional situation: a eulogy, graduation, retirement, award acceptance, or patriotic event, will use words creatively drawn by a skilled prose writer.
- **Political.** A speaker running for public office will make statements written by a team of political consultants.
- **Production.** A speaker at a "Big Tent" event (described in the next chapter) having to bridge to technical elements, such as, video clips or demonstrations, must use precise words to cue the production team.

In each of these cases, the speaker must rely on a hard copy of the speech—which sets up four potential pitfalls during its delivery.

- The speech appears and/or sounds "read" or "canned."
- The speaker loses Eye Connect with the audience.
- The voice becomes muffled by looking down at the text.
- The turning of the pages is distracting.

There are two solutions available:

- Teleprompter
- Vertical text

Teleprompter

This is the elaborate system traditionally used by U. S. Presidents in their State of the Union Addresses, the highest profile policy situation imaginable. Television news anchors and commentators use Teleprompter, too, but the system is also being deployed in many other circumstances, particularly in the business world.

In the State of the Union Address, the President stands on the dais, facing hundreds of legislators assembled in a Joint Session of Congress. Discreetly positioned at either side of the speaker's rostrum are two inconspicuous thin black rods supporting clear Plexiglas panels. These panels reflect the image of a concealed computer screen where the text of the speech scrolls by at a controlled rate. The angle of the panels makes the text visible only to the speaker and transparent to the audience, so it appears that the speaker is looking straight at the audience.

Thus, the Teleprompter system eliminates all four of the major pitfalls above, but it introduces three new ones, each of which requires a solution of its own.

- **Malfunction.** Mechanical devices can fail, so always have a hard copy as backup.

- **The Ping-Pong Effect.** As the speaker swings back and forth between the two panels, the repetitive pattern appears deliberate and unnatural. Ronald Reagan was a master of the Teleprompter game, moving back and forth between panels in smooth random swings, but he was a professional. As you read in Chapter 52, "Look, Ma, No Hands!" Mr. Reagan perfected his speaking techniques during his eight years as the host of the television series *GE Theater*.

  For those of us who are not Ronald Reagan, another solution is to use *four* Plexiglas panels, and to move among them in random sequence. Be sure to make these moves at logical points—when your voice drops to conclude a phrase.
- **Cadence.** The words you read will most likely have been written by another person whose cadence is different from yours. This could lead to a overly-deliberate or "canned" rhythm. The solution is to for you to preview the computer text and find your own cadence. Mark the ends of the phrases with backslashes. These marks will serve as your prompts to make your move to another Plexiglas panel.

Clearly, Teleprompter is a complex solution—and a costly one. For those on a tighter budget, there is another solution, good old reliable paper.

Vertical Text

When the text of a script is printed in conventional format, it causes readers to move their eyes in both the horizontal and vertical planes: across the line, down to the next line, and back to start the next line across. Please read the first two paragraphs of Abraham Lincoln's great Gettysburg Address below and feel your eye movements.

> *Four score and seven years ago our fathers brought forth on this continent, a new nation, conceived in Liberty, and dedicated to the proposition that all men are created equal.*
>
> *Now we are engaged in a great civil war, testing whether that nation, or any nation so conceived and so dedicated, can long endure. We are met on a great battle-field of that war. We have come to dedicate a portion of that field, as a final resting place for those who here gave their lives that that nation might live. It is altogether fitting and proper that we should do this.*

These moves are of no effect when you are reading from the printed page for your own enlightenment or enjoyment. But if you read this text to an audience, you would have to add two more vertical moves: up to look at the audience and back down at the page to go across the line. These additional movements make it difficult to keep track of your place on the page.

The solution is Vertical Text. Reset the tabs of your document to create a narrow column and reformat your script. Define each line as a piece of integral logic, mostly where the commas occur. These are the full phrases where you should drop your voice as you deliver them. Now read the reformatted Gettysburg Address below and note how your eye movements have diminished, making it easier to keep track of your place.

Four score and seven years ago

our fathers brought forth on this continent,

a new nation, conceived in Liberty,

and dedicated to the proposition that all men are created equal.

Now we are engaged in a great civil war,

testing whether that nation, or any nation

so conceived and so dedicated, can long endure.

We are met on a great battle-field of that war.

We have come to dedicate a portion of that field,

as a final resting place for those who here gave their lives

that that nation might live.

It is altogether fitting and proper that we should do this.

When you have reformatted your text, practice delivering it several times. As you become familiar with the content, you will pick up several lines each time you look down, and you will develop a fluid, comfortable, and varied rhythm.

Most important, you will be more connected with your audience.

69

Speaking to an Audience of a Thousand

The Big Tent

New product launches, trade shows, annual industry conferences, user group meetings, and similar large assemblages have become the rule rather than the exception in the business world today. Many companies, in their competitive drive to keep up with the Joneses, attempt to outdo each other with increasingly elaborate productions that rival the glitziest of Las Vegas spectaculars. As this tidal wave of extravaganzas sweeps forward, it threatens to drown the poor businessmen and -women who present at these events. Businesspeople are professionals at their jobs, but they are not performers.

How then to survive and thrive in this hypercharged environment? The short answer is to put yourself in the capable hands of the professional event planners, production supervisors, and technicians who manage these events. An entire subculture of experienced specialists has emerged to handle the increasing complexity of these mega-presentations. However, because these people are experts in the technical and production aspects, you must take responsibility for providing your own presentation script, slides, and demos.

When the size of the audience gets into the triple, quadruple, and even quintuple digits, the technicians begin to show up en masse bearing their forest of lights, miles of cables, wireless microphones, High Definition cameras, high resolution projectors, and gigantic screens. Note that I've pluralized "screens." As these events become more and more elaborate, the producers add more and more screens.

Jon Bromberg, the former Senior Director of Events at Microsoft, once produced a sales conference with 15 large screens arrayed across the stage.

Image Magnification

Typically, some of the screens display PowerPoint support slides, some of them thematic images related to the subject of a speech—monetary symbols, people at work, high tech gear, and so forth—and some of the screens show the presenter. The latter is known as Image Magnification, or as the technical people call it, "I-Mag."

In a venue where the audience can number in the thousands of people, the presenter appears as a mere speck. That is when the production team trains a battery of video cameras on the presenter and projects an enlarged close-up image onto one or more of the mammoth screens. And that is where the advice you read in Chapter 34, "Misdirection," no longer applies: When you are on I-Mag, do *not* turn to look at the screen where your PowerPoint slides are displayed.

If your magnified image turns on a giant screen, it appears as an abrupt disconnect from your audience. It also produces a distracting enlargement of your ear. Therefore, speak *only* to your audience.

How then to get the prompt that your slides usually provide? Your trusty production technicians will come to the rescue with large flat panel monitors that are strategically placed in front of you, at the foot of the stage. These monitors will be in your line of sight so that, when you glance away from your audience and down to the screen, your eye movement will be barely perceptible.

In this configuration, you get your prompt, you get your valuable pause, and your audience, their own eyes drawn up to the giant screens, will barely notice the minute shift of your eyes.

Actions, Speaking, and Words

These Big Tent events are theatrical productions, not presentations. Your role is subordinate to that of the technology; you function as a Master of Ceremonies rather than as a presenter.

In that role, you will be the bridge and/or the voice-over to a diverse array of technical elements far more complex than PowerPoint and decorative graphics. These elements can include prerecorded video clips, product demonstrations, satellite transmissions, or live closed circuit feeds, each of which must be integrated into the agenda with exacting precision. Therefore, all your actions and words will have to be tightly choreographed and scripted. To use terms familiar to professional performers, accept your role, learn your lines, hit your marks, and do it all on cue.

But keep in mind that *all* presentations are a series of person-to-person conversations. Even if you can only make direct Eye Connect with the people in the front rows, work your way around the large arena by speaking to points in the distance. Remember, too, to be conversational.

Best of all, you needn't fear committing an error. If you do, every man and woman in your audience will understand. They know that the Demo Demons (whom you will meet in the next chapter) can strike anyone, anywhere, or anytime. If the Demo Demons strike you, every member of your audience, who has been there and done that, will understand.

70

How to Beat the Demo Demons

Plan B and More

When companies bet the farm on new products, they often throw an elaborate launch party for industry influencers and the press. These Big Tent events are often centered on a demonstration of the precious newborn product. As anyone who has ever attended or, more to the point, participated in these events knows, the Demo Demons can strike suddenly, smiting the demo with a software crash, hardware jam, dropped Internet connection, frozen screen, power outage, blown lamp, dead battery, mouse malfunction, pop-up screen, or sound system failure.

Pure common sense, anticipation, preparation, and redundancy will avoid or counteract most of the physical and logistical pitfalls, but there is still the vital matter of showmanship: How do you *integrate* all the elements of your demo: the physical components, your voice and body language, and your all-important audience?

The Demo Demons may even give you a free pass and let you proceed without crashing, but you still have to make the demo go smoothly. After all, you are the surrogate for your audience; if you can't make your product work easily how can they?

Here are seven simple steps to make your next demo a success.

1. **Get out of the way.** When the demo starts, you become subordinate to the demo and are no longer the focus of the presentation. Shift your position to the periphery and make the demo the center of attention. Veteran actors get out of the way when

children and dogs are involved. The demo in a business presentation is the equivalent of the cute kid or the adorable puppy on stage. Step aside and let your demo be the focus.

2. **Become the Voice-Over narrator.** Take a lesson from how well-made documentaries do their narratives. The narrator is unseen, but tells a clear story. Be the VO for your demo, and make your narrative thorough. Don't be vague. Don't ad lib. Be as precise as a synchronized soundtrack.
3. **Say "you."** In their effort to appeal to the broadest possible customer base, most demos are impersonal tutorials that sound canned, and therefore, detached from the audience. By incorporating the word "you" into your narrative, you involve your audience as if they are participants in the demo and, by implication, as buyers of the product you are demonstrating.
4. **Make Eye Connect via the demo.** Look at your own demo or at the projection screen displaying your demo. Don't try to engage with your audience's eyes as you demonstrate. They are focused on the demo itself or on the screen, and not on you. If you turn to face them, they become conflicted about where to look. This is the same dynamic you read about in Chapter 34, "Misdirection." Take the same point of view as your audience. Watch the demo evolve as if you were an end user.
5. **Pause for action.** Allow the demo to speak for itself, especially when many actions are occurring at the same time. Pause at such points and allow the action to run its course. Silence is golden and actions speak louder than words.
6. **Use verbal navigation.** This is an extension of the same technique you learned in Chapter 45, "No More Mind-Numbing Number Slides." Navigate your audience's eyes with your words. Direct their eyes by describing what they are seeing in the demo. Reference colors, objects, position, and direction. Reference top and bottom, center and sides. Reference left and right, too, but be sure that you make it clear *which* left or right. Don't force your audience to do the navigation. A simple fail-safe is to orient yourself with same point of view as your audience.
7. **Learn Sullivan's Law.** Sullivan's Law is a corollary of Murphy's Law, which is, "Anything that can go wrong will." Sullivan's Law is "Murphy was an optimist."

Develop a Plan B for what you will do and say if the Demo Demons strike.

71

Bring Your Panel Discussion to Life

How to Herd Cats

The bestselling book *Death by Meeting* could readily have a sequel called *Death by Panel Discussion.* If you've attended just one conference in your life, whether it was business, professional, academic, or even social, you must be painfully aware of the soporific effect of panel discussions. The fault, however, lies not with the panelists, but most often with the moderator, who inevitably allows the panelists to pontificate, dawdle, or wander off into black holes, causing the audience to nod off.

At some point in your life, you may be asked to chair a panel. To bring your discussion to life, incorporate these ten best practices.

1. **Invite conflict.** Conflict is drama, Aristotle 101. In your preparation for the panel discussion, find the hot buttons in the major subject, as well as in the position of each panelist. Select panelists with wide diversity and/or differences of opinion. As moderator, throw the conflict into the mix.
2. **Make a wave.** Pose one question and go down the line, asking each panelist to respond.
3. **Counter-punch.** Listen for differing positions in the panelists' responses and confront one panelist to defend or rebut the other.
4. **Gather questions from the audience in advance.** The operative word is "advance," which enables you to filter only the relevant questions. The usual opening of the floor to questions from the audience gives the impression of being warm and fuzzy but, more often than not, deteriorates into digressions

that spiral off into still deeper black holes. Prescreening the questions keeps your panel on track.

5. **Manage the time.** And do it with strict observance of the Less Is More principle. Keep every statement or exchange by every panelist to a maximum of 2 to 3 minutes. Limit opening and closing statements to 90 seconds. Ask the conference manager to provide a large countdown clock or two, visible to both the panelists and the audience. This gives the panel the look and feel of a well-produced television debate. More important, it moves the discussion along at a brisk pace, a rare phenomenon in most panels.
6. **Suit up and show up.** A panel discussion is a form of presentation, so it is important for the panelists to present at their best. The same is true for you as the moderator. Establish a dress code with coats and ties for the men and smart clothing for the women. Forget the current custom of open collar shirts for men; even if the conference takes place at a casual resort. Open collar shirts inevitably scrunch up beneath the jacket collars, making the wearers look like mutant turtles. Ask your panelists to stand out. Forget about "When in Rome..."
7. **Seat the panelists on stools.** The usual panel living room set, imitative of late night television talk shows, is often furnished with comfortable sofas and armchairs. This looks cozy, but causes the panelists to slump deep into the upholstery, scrunching their chests, constricting their lungs, thus producing inaudible mumbles, indecipherable even by clip-on microphones. The slumping also occasionally produces loud, disembodied thumping, whooshing, or scraping noises on those same microphones. To make matters still worse, cushy furniture takes the panelists' arms out of service, making them appear inanimate.

   Stools force the panelists to sit up straight, which, in turn, frees their arms to gesture. Sitting up also opens panelists' chests to give full range to their voices—just like Frank Sinatra and all the other professional singers who give full range to their voices by standing up straight when they sing.
8. **Be the apex of a triangle.** Most moderators sit in profile at either end of the panel, while the panelists sit full front facing the audience. This arrangement diminishes the role of the moderator and allows the panelists to grandstand. Forming a triangle, with the moderator at the apex and the panelists along the legs, gives greater prominence and control to the moderator.

9. **Plan a surprise.** Introduce an unexpected element during the discussion: an unannounced guest, a video clip, a photograph, a quote, or a news story. Then ask the panelists to respond.
10. **Keep it moving and mix it up.** The moderator can provide CPR to the panel by maintaining a brisk pace and by bouncing the focus from one panelist to another often.

Variety is the spice of life—and panels. Bring life to your panel discussions.

72

Mark Your Accent

Eliza Doolittle Is a Myth

The inexorable march of globalization in business has created a vast landscape of diverse workforces of diverse origins, all working together but speaking in a Babel of tongues. In their desire to communicate effectively, transplanted workers seek to learn the language of their adopted homes in classes and books, from CDs, computers, and online programs. These expatriates also seek to speak their newly acquired language more clearly, which has created a surge in a veteran niche industry called "Accent Reduction."

An Internet search of that term yields more than a million entries posted by scores of present-day versions of Henry Higgins offering their services to multiple variations of Eliza Doolittle to help them to become not only fair ladies and gentlemen, but to pronounce their words free of native accents that can mask their meanings.

Remember, however, that the source for *My Fair Lady* was George Bernard Shaw's *Pygmalion*, and that the source for *Pygmalion* was mythology. In that legendary tale, Pygmalion, a Roman sculptor, worked for years to carve a stone statue of his ideal woman. Pygmalion became so enamored of the statue he named Galatea, that he prayed for the Gods bring her to life and, magically, they did.

The operative word in myths is "magically." The operative word in reality is, to quote Eliza Doolittle, "not bloody likely." "Reduction" is a more attainable goal than the *elimination* that Eliza achieved, but elimination is nearly impossible in the nonfiction world for one simple reason: Speaking is a habit practiced since infancy, and habits of that duration are extremely difficult to break.

Does that leave those of you who are striving to adapt to your new environment hopelessly stuck at Square One, resigned to pronounce your English words marked, if not obscured by your native accent? Not quite, but the hard, cold fact of a deeply imbedded—and therefore subconscious—speech pattern means that you must set realistic goals and choose the best path to achieve them. To help you find your path, let me share with you how I learned to speak Spanish.

Spanish Lessons

My first acquaintance with the native tongue of Miguel Cervantes was in the New York City public school system where, like most other U.S. school systems, the approach to language lessons is to drill students with copious lists of the days of the week, months of the year, cardinal and ordinal numbers, common nouns, and verb conjugations. My second experience, in a Spanish course in college, was the same.

By the time I was ready to fly solo in public, my brain went into calisthenics mode: Every time I searched for a Spanish word, my mind went through a prescribed step drill of the entire list of nouns and the conjugations of each verb. What came tumbling out of my mouth was more spastic than Spanish.

This inefficient system continued to haunt and embarrass me into adult life. Every time I tried to speak Spanish—on vacations in the Caribbean, in restaurants and commerce with Hispanic workers in New York, and in a brief professional foray in Spain—I felt as if was stumbling down a long flight of stairs. But then in 1983, I had a long professional foray that radically changed the system.

In what was to be my last job in television before founding Power Presentations, I was hired to produce a drama series for Televisa, the Mexican media conglomerate. I moved, lock, stock and barrel, to Mexico City to fulfill a year-long contract. Because the series was to be produced in English, the position did not require me to speak Spanish, but it did require me to communicate with people who spoke no English.

For that entire year, I had to speak Spanish in order to eat, drink, shop, and travel. Before I knew it, the step drill in my mind had vanished. Within three months, I became fluent enough to speak Spanish in production conferences with my Mexican colleagues and in script conferences with an Argentine writer named Norberto Vieyra, who spoke no English.

The only impediment to our story discussions was *accent*. Argentine Spanish is pronounced very differently from Mexican Spanish, and Norberto's accent was as difficult for me, as mine—in Spanish—was for him. My newfound fluency in verbiage had been accompanied by a fluency in accent. Norberto had difficulty understanding me because I sounded Mexican!

Whither the fluency? I had taken no lessons, read no books, listened to no tapes. But I had listened to Spanish: on television and radio, in the Televisa offices, on the streets, and in the shops and restaurants of Mexico City.

Ears Versus Eyes

The operative word in learning languages, therefore, is "listen." I finally learned to speak Spanish by hearing it spoken and then speaking it myself. All it took for me to learn was to listen to new words and then to repeat them. If that process sounds familiar, it should, for it is just how all human beings learn to speak all languages—as children. Children listen to their parents and then speak. Period. No books, no lists, no memorization, no conjugation, no thinking. The brain and the eyes are not involved in the process, only the ears and the mouth.

Listen and speak. Hear the words and imitate the sounds. The accent comes with the territory.

Berlitz, the venerable international language school has long offered a program called Total Immersion, which involves organized travel to a destination where the desired language is spoken; to Mexico for Spanish or to France for French. Once there, the students, shepherded by Berlitz instructors, spend one to three weeks living as a native, speaking only the target language. My Mexican experience was, in its own way, total immersion—without the supervised instruction.

If you want to reduce your accent, develop your own form of immersion. Listen to English on radio and television or to podcasts on the Internet. Hear the words and imitate the sounds. Open your mind and your ears to the English you encounter in your daily life and repeat what you hear. Before long, your English will sound more like that of a native rather than filtered through the pronunciation pattern of your original language. Your accent will never vanish, but you will sound clearer.

Repetition plus imitation equals Verbalization, the rehearsal technique you read about in Chapter 46, "Eight Presentations a Day." Verbalization is one of the most powerful tools you can use to improve the fluidity of your presentation; you can also use it to improve the clarity of your speech—and become a fair lady or gentleman.

73

How to Interview Like a Television Anchorperson

Seven Easy Steps

The Aspen Ideas Festival is a high-profile annual summer conference at which some of the best minds in diverse fields (among them culture, economics, politics, science, and education) come to share their concepts with audiences at the Festival. To avoid having these big thinkers give tedious unilateral lectures, most of the sessions are interviews conducted by respected journalists or peers.

The interview approach has also caught on at investment conferences where, rather than having senior executives of presenting companies deliver one-way PowerPoint presentations, they are interviewed by financial professionals in what are called "fireside chats." The interview approach has long been a staple of large-scale business conferences where prominent industry thought leaders are interviewed by select individuals.

Someday, you might be selected to be that individual.

Unfortunately, most of the individuals who are given this responsibility are so awed by their guest's prominence, they ask only innocuous questions, leaving the guest no choice but to deliver a tedious monologue.

This is not to say that, if you become an interviewer, you should swing to the other end of the scale and attack your guest with the slings and arrows of the late Mike Wallace or the present Bill O'Reilly. Instead make your role model Charlie Rose or Barbara Walters, both of whom have the unique ability to draw interesting stories out of their guests. While a businessperson can't expect to match Mr. Rose's or Ms. Walters's professional acumen, it is possible to lead your guest through an interesting session with these seven easy steps:

1. **Set the context.** Most interviewers fail to remember the First Commandment of presentations: *Tell 'em what you're gonna tell 'em.* Right at the beginning of your interview, establish a roadmap of what subjects you intend to cover.
2. **Navigate.** Most interviews bounce around randomly. To avoid such rambling, the interviewer must establish a structure and then keep it on that track. One of the simplest and clearest structures is to follow a timeline, ticking off the milestones by dates. Another is to organize the interview in a series of numbered steps. As you read in Chapter 26, "David Letterman's Top 10," this simple technique brings order to what could otherwise be chaos. Lead your subject into the details of a time period or a number, and then move them to the next period or number in the sequence.
3. **Be transparent.** The influence of television's current cult of belligerent personality interviewers has driven many nonprofessional interviewers to insert their own points of view, and some to do so aggressively. Wrong!

   A *Wall Street Journal* article written by Peter Funt recounted "a recent interview Lawrence O'Donnell conducted on MSNBC with the filmmaker and activist Michael Moore. Mr. Moore spoke a total of 1,034 words, while Mr. O'Donnell—whose job, after all, was to ask questions—spoke almost as many: 900. The host was so intent on both asking and answering questions that at one point Mr. Moore said jokingly, 'Thanks, Lawrence, for coming on the show tonight.'"

   As the son of Allen Funt, the creator of the famous "Candid Camera" television series, Peter Funt writes with authority about interviewing. He went on to say that "The best interviewers do their homework, put their own opinions aside, keep questions brief, and listen closely for possible follow-ups."[1]

   Charlie Rose and Barbara Walters do just that. Their talent is their ability to interview and yet be transparent at the same time. Mr. Rose described his interviewing techniques during his own interview at the 2012 Aspen Ideas Festival:

It is often said that the most important question is, "Yes, but.." But even today, the most important questions are..."Why?" and "Why not?" "What were your choices?"

I'm always conscious of the sense that, if I'm looking and talking to someone, you always want to know who their heroes are, and who's influenced them.[2]

The subject of any interview is the interviewee who has achieved whatever he or she has accomplished to get them invited to be interviewed in the first place. You can gently nudge your guest toward salient points along the way, but let the star have the spotlight.

4. **Summarize.** Interviewees, who have lived every moment of their own lives, tend to get into granular details during the interviews. Even articulate subjects must be netted out. At salient milestones, recap the high points of what your subject has said. Whenever possible, go beyond summarizing to provide perspective by adding information about the subject that you have researched in advance.
5. **Involve the audience.** An interview does not exist in a bubble. Although silent, the audience is a third party in any interview. From time to time, turn to your audience for a reaction. Get them to respond with applause or a show of hands or, if there are microphones available, a few verbal comments—with an emphasis on "a few."
6. **Cross-reference.** The interview does not exist as a floating point in time, particularly if it is part of a conference that has other events such as panel discussions or speeches. You can make references to related subjects or events before or after the interview or to points made earlier in the interview itself. Do both, and do it often. Weave a web of linkages.
7. **Draw best practices.** The anecdotal nature of interviews makes the point of view linear. The interviewer can add depth and breadth by drawing out lessons learned. At pivotal points, ask your guest what worked well and what he or she would have done differently. From that, your audience will learn what to do, and what not to do. Give your audience extra added value beyond the view of the celebrity.

Following these seven steps might not get you on national or even cable television, but they will keep the tedium out of your interview and, more important, earn you the gratitude of your audience.

74

Ten Best Practices for the IPO Road Show

If I can make it there, I'll make it anywhere.

—"New York, New York,"
Lyrics by Fred Ebb

In the Introduction, you read that, as the coach of the IPO road shows of nearly 600 companies, I applied the same techniques as I did for hundreds of other companies to develop thousands of presentations to raise private financing, sell products, form partnerships, or gain approval for internal projects.

While very few people get the opportunity to make a presentation that seeks to raise north of 150 million dollars, in the ten best practices for IPO road shows that follow, I'm confident you'll see aspects that resonate with any presentation—and with the techniques throughout this book. As you read these practices, please relate them to your own presentation circumstances.

1. **The NetRoadshow Factor.** In 2005, the Securities and Exchange Commission mandated, in the interest of full disclosure, that companies offering stock for the first time must make their road show presentation available to the public online. Since then, every company makes a video recording of the

management team delivering their pitch and posts it on the NetRoadshow/RetailRoadshow site along with the slideshow that accompanies their narrative, as in Figure 74.1.

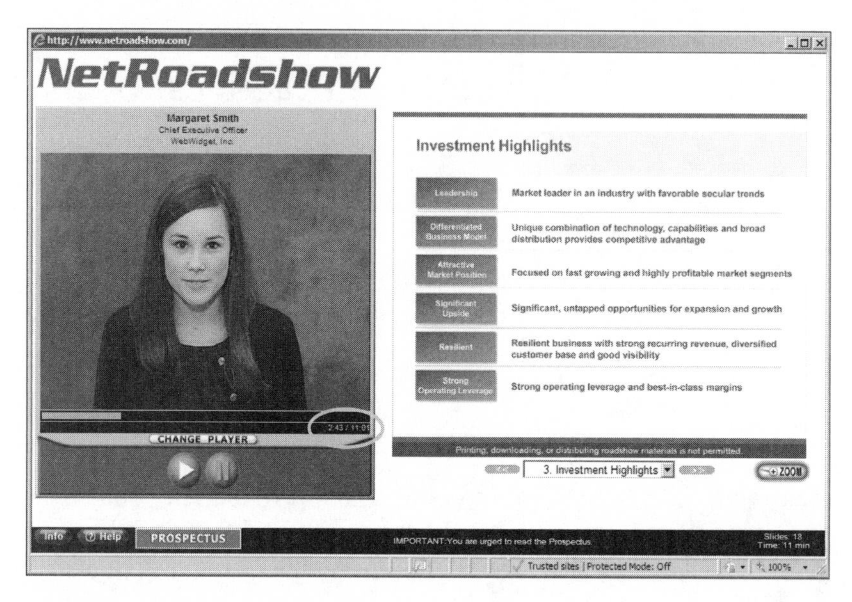

Figure 74.1 NetRoadshow Web page

Despite this broad access, the company's senior management team goes on the road for about two weeks, during which they visit potential investors in about a dozen cities across the country for about 30 or 40 meetings a week for a total of 80 or more iterations—just as you read in Chapter 27, "Illusion of the First Time." The reason for this grueling tour is that no investor will make a decision to buy up to a 10% tranche of the offering (doing the math, that could be a $15 million decision) based on a canned presentation alone. Investors want to meet the executives in person, press the flesh, look them in the eye, and interact with them directly.

Investors have come to rely on the NetRoadshow presentation to get the basics of a company's story, but they still want to meet the management in person. As a result, many of the meetings are not presentations, but intense Q&A sessions.

But meet they must, and presentation counts.

2. **Timing.** If you look just below the image of the presenter in Figure 74.1, you'll see the elapsed as well as the total running time of the presentation. Therefore, every investor who clicks onto the site knows just how long the presentation will last—and the site does not allow fast forward. When you consider that investors who, in the course of their daily jobs, spend most of their time making split-second decisions on trades, and who are likely to be reading and sending emails, instant messages, and tweets while accessing NetRoadshow, you'll know that this is an audience with very little patience.

   Most road shows try to come in under 30 minutes. Pity the poor company that posts a longer running time.

   Less is More has never had a better proof point.

3. **New audience, new benefits.** Until the time of the public offering, most presentations that most companies deliver are to customers. Customers have different benefits than do investors, and so presenters must make that switch. A case study of an IPO from *Presenting to Win* illustrates this point:

   Jim Bixby was the CEO of Brooktree, a company that made and sold custom-designed integrated circuits used by electronics manufacturers. (It was later acquired by Conexant Systems, Inc.) In preparation for Brooktree's IPO, Jim rehearsed his road show with me. I role played a money manager at Fidelity.

   During the product portion of Jim's presentation, he held up a large, thick manual and said, "This is our product catalog. No other company in the industry has as many products in its catalog as we do."

   Jim set down the catalog and was about to move on to the next topic when I raised my hand and said, "Why should I care?"

   With barely a pause, Jim raised the catalog again and replied, "With this depth of product, we protect our revenue stream against cyclical variations."

   That was the right benefit for the right—investor—audience. Tailor every benefit for every audience.

4. **Competition.** Any company in any market has competitors with whom they are engaged in a mortal struggle—or the market would be worthless. However, investors, seeing the worth in a market, sometimes buy stock in several competing companies. Therefore, when discussing the competitive landscape, lest you make the investor nervous, it would be unwise to bash the competition. Instead, downshift to compare and contrast the entire competitive landscape.

5. **Web Animation.** The animation feature in PowerPoint, although abused by some presenters, can be an aid to telling a story. At its most basic, animation can be used to present information in incremental, digestible bites. In order to ensure that the NetRoadshow presentation can be accessed by every investor regardless of desktop configuration, the NetRoadshow platform allows for limited animation. As a result, presenters may want to make a separate version of the slideshow using redundant slides to create the additive build effect.

   The additive build effect makes images easier for audiences to process.

6. **Flow Structure.** An IPO road show, just like *any* presentation, must have a clear and logical story arc. The structure of choice for most road shows is the same—for a simple reason: to parallel the Prospectus. The centerpiece of this carefully wrought document is a "Business" section divided into two major subsections: "Industry Overview," which is written as an opportunity, and "Company Solution," which is written as the leverage of the opportunity.

   I was privileged to coach the road show of Cisco Systems when they went public in 1990. Here from *Presenting to Win* is their story arc:

   The Cisco team began by describing the shift in computing from mainframes to PCs. They then moved on to delineate the rapid growth of local area networks and wide area networks (LANs and WANs), and recent improvements in technology that brought significant increases in speed, bandwidth, and power. They ended this cluster with a look at the anticipated shift in business from enterprise-centered to remote-based computing. All of those trends, taken together, represented an opportunity.

   Next, they talked about how their new device, called a router, could internetwork all networks. They explained how Cisco manufactured the router, how they serviced it, how they sold it through channels and strategic relationships, and where they intended to go with the router in the future. All of these facts, taken together, represented Cisco's leverage of the opportunity.

   In today's high-speed, PowerPoint-driven world, presenters often hurriedly cobble together a disparate set of slides and proceed to deliver them in a sequence that only they understand. Take the time to create your story with a clear, logical arc that makes it easy for your audience to follow—or they will make it hard for you.

7. **Verbalization.** In Chapter 46, "Eight Presentations a Day," you read about the value that Noland Granberry, the CFO of Silicon Image, Inc., experienced from speaking his presentation aloud multiple times. Please note that Mr. Granberry is an American who was speaking in his native language. In today's globalized world, many speakers must present in a second language. Verbalization is even more helpful to them.

   Take the case of a large Israeli technology manufacturing company whose IPO road show I coached. I spent several days working with them at their plant outside Tel Aviv, but before returning to the United States, I urged them to Verbalize on their own. They did so with a vengeance, presenting to several different groups of employees: in the cafeteria, small conference rooms, training classrooms, and large meeting rooms.

   When they started their road show, I went to see them present at a luncheon meeting in San Francisco. They had presented in Los Angeles that morning, but due to the usual San Francisco fog, their flight was delayed. You can image the mood in the room among those characteristically impatient investors. However, when the Israeli team arrived, they were so well prepared, they presented with complete cool, calm control. And they were speaking English as a second language!

   Verbalization works.

8. **Team building.** The Israeli team above experienced two other benefits from the multiple iterations of their presentation to multiple groups of employees:

   - The employees felt involved.
   - The employees provided unique insights on the company's story.

   Rehearse your presentations to your colleagues to gain perspective.

9. **Assertive Language.** In Chapter 11, "Meaningful Words," you read how the conditional mood, used to avoid forward-looking statements, weakens narrative language. That problem is heightened in IPOs because the senior management team that delivers the road show spends many hours over many months in drafting sessions for the Prospectus. Those sessions are led by attorneys whose customary cautionary views produce a document composed almost exclusively in the conditional mood. That mood spills over into the road show.

Far be it from me to prompt CEOs and CFOs to switch to the declarative mood and risk making those dreaded forward-looking statements. But I do urge them—and *you*—to replace "We believe..." "We think..." and "We feel... with "We're confident...."

10. **Make direct references to each audience.** Here, I reiterate the importance of "the illusion of the first time," the subject of Chapter 27. I reiterate it because most IPO road show CEOs and CFOs—and most presenters—neglect to do so.

    I saved this reiteration for last because, of all the techniques in this book, this one provides the biggest bang for the buck; and yet, in the heat of battle, it is implemented the least.

    Use it or lose it.

75

Cicero: Peroration

Timeless and Borderless

In the Introduction, I promised to show you that effective presentations share the same essential elements required in many other forms of communication, and that all these elements traverse time and place to make them universal:

- Telling a clear, logical story
- Designing simple, effective graphics
- Delivering with confidence and authority
- Handling challenging questions effectively

I hope that I have fulfilled my promise. I also promised to conclude with a bookend by the great Roman orator, Marcus Tullius Cicero. I've chosen an excerpt from his essay, "On the Character of the Orator," to serve as the peroration because it perfectly summarizes the principal best practices in this book —and he identified them more than two thousand years ago:

> *Eloquence, in fact, requires many things: a wide knowledge of very many subjects (verbal fluency without this being worthless and even ridiculous), a style, too, carefully formed not merely by selection, but by arrangement of words, and a thorough familiarity with all the feelings which nature has given to man, because the whole force and art of the orator must be put forth in allaying or exciting the emotions of his audience.*

Further than this it requires a certain play of humour and wit, liberal culture, a readiness and brevity in reply and attack, combined with a nice delicacy and refinement of manner. It require also an acquaintance withal history, and a store of instances, no can it dispense with a knowledge of the statute-books and all civi law.

I need hardly add, I presume, any remarks on mere delivery. Thi must be combined with appropriate movement of the body, gestures, looks, and modulation and variety of tone. How importan this is in itself may be seen from the insignificant art of the acto and the procedure of the stage; for though all actors pay grea attention to the due management of their features, voice, and gestures, it is a matter of common notoriety how few there are, o have been, whom we can watch without discomfort.

One word I must add on memory, the treasure-house of all knowledge. Unless the orator calls in the aid of memory to retain the matter and the words with which thought and study have furnished him, all his other merits, however brilliant, we know wil lose their effect.

We may therefore well cease to wonder why it is that real orators are so few, seeing that eloquence depends on a combination of accomplishments, in each one of which it is no slight matter to achieve success.[1]

4. James Collins, "The Plot Escapes Me," *New*
5. Guy Kawasaki, *Enchantment* (Portfolio, 2011

Chapter 5

1. Federico Fellini, *Fellini on Fellini* (Da Capo
2. Casey Schwartz, "ADHD's Upside is Creati *Beast*, February 8, 2011.

Chapter 6

1. PBS.org, *American Masters*, "Woody Allen: July 21, 2011. www.pbs.org/wnet/americ documentary/about-the-film/1865/.
2. Mindjet, www.mindjet.com.
3. Microsoft Support, "A Comparison of Visio http://support.microsoft.com/kb/896660.
4. A. O. Scott, "When in Rome, Still an Anxiou *Times*, Movies, June 21, 2012.
5. PBS.org, *op. cit.*

Chapter 7

1. "Existentialism," Wordnik, www.wordnik.co
2. Basic Funny Quotes, "Woody Allen," www.ba
3. Mick LaSalle, "Review: Woody Allen's Mea *cisco Chronicle*, October 1, 2010.
4. Stephen R. Covey, *The 7 Habits of Highly E* sion of Simon & Shuster, 1989, 2004), p. 95.

Chapter 8

1. Terry Teachout, "Manipulated Puppets, M *nal*, April 15, 2011.
2. A. O. Scott and Manohla Dargis, "Better t *York Times*, March 24, 2011.

Chapter 9

1. Better Presenting, Annual Conference, ww

Chapter 12

1. A. O. Scott, "A Simple Prescription for Sup *less*, *The New York Times*, March 17, 2011.

Chapter 13

1. Clive Thompson, "Clive Thompson on Capturing Aha Moments," *Wired Magazine*, May 31, 2011.
2. Charles McGrath, "Nicholson Baker: The Mad Scientist of Smut," *New York Times Magazine*, August 4, 2011.
3. Clive Thompson, *op. cit*.

Chapter 14

Theme image, British Prime Minister Sir Winston Churchill addresses joint session of Congress, January 1942, public domain

1. The Churchill Centre and Museum at the Churchill War Rooms, London, "Quotes FAQ," www.winstonchurchill.org/learn/speeches/quotations/quotes-faq. The full speech is contained in *The Unrelenting Struggle* (London: Cassell and Boston: Little Brown, 1942) and is found on pages 274-76 of the English edition. It may also be found in *The Complete Speeches of Winston S. Churchill*, edited by Robert Rhodes James (New York: Bowker and London: Chelsea House, 1974).
2. Mardy Grothe, *Neverisms: A Quotation Lover's Guide to Things You Should Never Do, Never Say, or Never Forget* (Collins Reference, 2011).
3. Erin McKean, "How's 'Never' for You?" Boston.com, June 12, 2011.
4. Dave Sterry and Arielle Eckstut, eds., *Satchel Sez: The Wit, Wisdom, and World of LeRoy "Satchel" Paige* (Three River Press, 2001).

Chapter 15

1. IMDb, *Casablanca* (1942), "Memorable Quote for *Casablanca*," www.imdb.com/title/tt0034583/quotes.
2. David Brody, "Newt Gingrich Tells the Brody File He 'Felt Compelled to Seek God's Forgiveness,'" The Brody File, CBN News, Blogs, March 8, 2011.
3. John M. Broder, "Familiar Fallback for Officials: 'Mistakes Were Made,'" *The New York Times*, March 14, 2007.

Chapter 16

1. Cassell Bryan-Lowe and Russell Adams, "Dow Jones CEO Resigns Over Scandal," *Wall Street Journal*, July 16, 2011.
2. John F. Burns, "A Day of Apologies for the Murdochs, and of New Questions for Cameron," *New York Times*, July 16, 2011.
3. Dan Sabbagh and Josh Halliday, "Rupert Murdoch 'Not Fit' to Lead Major International Company, MPs Conclude," *The Guardian*, May 1, 2012.
4. Paul Sonne and Jeanne Whalen, "News Corp. Blasted in U.K.," *Wall Street Journal*, May 1, 2012.

Chapter 17

Theme image, Bill Clinton, official presidential portrait, www.whitehouse.gov, public domain

1. Robert Yoon, "Clinton Earns $65 Million in Speaking Fees as Private Citizen," CNN Political Ticker blog, June 29, 2010.
2. Peggy Noonan, "The Life of the Party," *Wall Street Journal*, August 4, 2012.
3. Chris Lehourites, "Clinton, Freeman Not Dazzling in Cup Pitch," *NBC Sports*, December 1, 2010.
4. December 1, 2010: USA 2022 Bid Presentation, YouTube.com, www.youtube.com/watch?v=Oe6yMPeU0DE&feature=related.
5. Steve Kelley, "Qatar? Really? FIFA Awards 2022 World Cup to Soccer-Poor, Oil-Rich Qatar," *Seattle Times*, December 2, 2010.
6. Matt Scott, "Mike Lee, the Englishman Behind Qatar's World Cup Success," *The Guardian*, December 2, 2010.

Chapter 18

Theme image, John Doerr, Kleiner, Perkins, Caufield & Byers

Chapter 19

Theme image, Vinod Khosla, Khosla Ventures

Chapter 20

1. John Jay College of Criminal Justice, "Basic Outlining," www.lib.jjay.cuny.edu/research/outlining.html.
2. Rachel Emma Silverman, "Doodling for Dollars," *Wall Street Journal* Online, April 24, 2012.
3. Andy Fixmer, "Encyclopaedia Britannica Ends 244-Year-Old Print Edition," Bloomberg.com, March 14, 2012.
4. Walter S. Mossberg, "Encyclopaedia Britannica Now Fits into an App," *Wall Street Journal,* September 29, 2011.

Chapter 21

1. Joe Dator, "Enough about me but nothing about you just yet," *New Yorker* Cartoon, May 4, 2009.
2. Sherry Turkle, "The Flight from Conversation," *New York Times,* April 21, 2012.
3. James W. Pennebaker, *The Secret Life of Pronouns* (Bloomsbury Press, 2011), p. 3.
4. Tara Parker-Pope, "Can Romance Be Reduced to Pronouns," *New York Times,* The Well Column, October 27, 2011.

Chapter 22

Figure 22.1, courtesy of Social Sciences – Psychological and Cognitive Sciences – Biological Sciences – Neuroscience, Diana I. Tamir and Jason P. Mitchell, *Disclosing information about the self is intrinsically rewarding,* PNAS 2012 109 (21) 8038-8043; published ahead of print May 7, 2012, doi:10.1073/pnas.1202129109.

1. Diana Tamir and Jason Mitchell, "Disclosing Information about the Self Is Intrinsically Rewarding," Proceedings of the National Academy of Sciences of the United States of America, May 7, 2012, www.pnas.org/search?fulltext=tamir+and+mitchell&submit=yes&go.x=14&go.y=13.
2. Robert Lee Hotz, "Science Reveals Why We Brag so Much," *Wall Street Journal*, May 7, 2012.
3. Greg Behrendt, *He's Just Not That Into You* (Simon Spotlight Entertainment, 2004).
4. Clayton M. Christensen, Scott Cook, and Taddy Hall, "Marketing Malpractice: The Cause and the Cure," *Harvard Business Review,* December 1, 2005.

Chapter 23

1. "Barry Goldwater Quotes," Brainy Quote.com, www.brainyquote.com/quotes/authors/b/barry_goldwater.html.
2. "Bruce Erik Kaplan," Simon & Schuster, Authors, http://authors.simonandschuster.com/Bruce-Eric-Kaplan/1348342.
3. Bruce Erik Kaplan, First I want to give you an overview of what I will tell you over and over again during the entire presentation, *New Yorker* cartoon, April 9, 2012.
4. Geoff Dyer, "An Academic Author's Unintentional Masterpiece," *New York Times,* July 22, 2011.

Chapter 24

1. John Irving, "The 'Cider House' Writing Rules," *Wall Street Journal,* May 4, 2012.
2. William Safire, "Gifts of Gab," *New York Times Magazine,* "On Language," December 12, 2004.
3. Jhumpa Lahiri, "My Life's Sentences," *New York Times,* Opinionator blogs, March 17, 2012.

Chapter 25

1. "Audience IPO Prices above Range for Thursday Debut," *Wall Street Journal,* May 10, 2012.
2. Lynn Cowan, "WageWorks, Audience Rise in Debut," *Wall Street Journal,* May 10, 2012.
3. "Diabetes Statistics," American Diabetes Association, www.diabetes.org/diabetes-basics/diabetes-statistics/.

Chapter 26

1. *Late Show with David Letterman,* CBS.com, www.cbs.com/late_night/late_show/top_ten/.
2. Politico, www.politico.com/.

Chapter 27

Theme image, "Sherlock Holmes," William Gillette, www.loc.gov/pictures/resource/var.1362/, public domain

1. William Gillette, *The Illusion of the First Time in Acting* (New York: Dramatic Museum of Columbia University, 1915), p. 43.

Chapter 28

1. Harvard Business School, "Academics," www.hbs.edu/mba/academics/casemethod.html.
2. Annie Murphy Paul, "Your Brain on Fiction," *New York Times,* March 17, 2012.
3. Barry Newman, "What Is an A-Hed?," *Wall Street Journal,* November 3, 2010.

Chapter 29

Theme Image, Barack Obama, official presidential portrait, www.whitehouse.gov, public domain

Theme Image, Ronald Reagan, courtesy of the Ronald Reagan Library, public domain

1. Barack Obama, *The Audacity of Hope: Thoughts on Reclaiming the American Dream* (Crown, 2006), p. 31.
2. Pham Sherise, "Obama Reading About Reagan," ABC News, December 23, 2010.
3. Lou Cannon, *President Reagan: The Role of a Lifetime* (Public Affairs, 2000), p. 20.
4. "Remarks by the President at a Memorial Service for the Victims of the Shooting in Tucson, Arizona," The White House, Office of the Press Secretary, January 12, 2011, www.whitehouse.gov/the-press-office/2011/01/12/remarks-president-barack-obama-memorial-service-victims-shooting-tucson.
5. Peggy Noonan, "Obama Rises to the Challenge," *Wall Street Journal*, January 15, 2011.

Chapter 30

Theme image, Aristotle, http://starchild.gsfc.nasa.gov/Images/StarChild/universe_level2/aristotle.gif, public domain

1. Mortimer J. Adler, *How to Speak How to Listen* (Scribner, 1983), p. 29.

2. Philip Delves Broughton, *The Art of the Sale: Learning from the Masters about the Business of Life* (Penguin Press HC, 2012), p. 2.
3. L. Gordon Crovitz, "Have We Got a Review for You," *Wall Street Journal*, April 27, 2012, p. A13.

Chapter 31

Theme image, Vinod Khosla, Khosla Ventures

1. AT-A-GLANCE, "About Us," www.ataglance.com/ataglancestore/common/static.jsp?pageId=AboutUs

Chapter 32

1. The Co-Creativity Institute, "Staff and Resources," www.cocreativity.com/staff.html.
2. Christopher M. Barlow, "Thinking More Effectively about Deliberate Innovation," *Howe School Alliance for Technology Management*, Spring 2007, Volume 11, Issue 1.
3. Ron Gluckman, "Skyscraper, Moneymaker," *Wall Street Journal*, April 12, 2011.

Chapter 34

1. Joe Morgenstern, "Towering 'Tree,' Out on a Limb," *Wall Street Journal*, May 27, 2011.

Chapter 35

Theme image, President Barack Obama, www.whitehouse.gov, public domain

1. Abby Phillip, "After Leak, Obama Reconsiders," Politico44, January 25, 2011.

Chapter 36

1. The Criterion Collection, *Ikiru*, www.criterion.com/films/353-ikiru.

Chapter 37

1. Matt Zoller Seitz, "How Hollywood Killed the Movie Stunt," *Salon*, November 12, 2010.

Chapter 38

Theme image, reprinted by permission of Anti-PowerPoint Party

1. Edward Tufte, "PowerPoint Is Evil: Power Corrupts. PowerPoint Corrupts Absolutely," *Wired Magazine*, September 2003.

2. Anti-PowerPoint Party, www.anti-powerpoint-party.com/.
3. Lucy Kellaway, "Anti-PowerPoint revolutionaries unite," *Financial Times*, July 17, 2011.
4. Michael Baldwin, "PowerPoint can pack a punch—in the hands of a wizard," FT Letters to Editors, *Financial Times*, July 22, 2011.

Chapter 39

1. Paul Shaw, *Helvetica and the New York City Subway System* (The MIT Press), February 11, 2011.

Chapter 40

1. The Quotations Page, Quotation #26229 from Classic Quotes, Henry David Thoreau, www.quotationspage.com/quote/26229.html.

Chapter 41

Theme image, Illuminated D, www.freepik.com/free-photo/illuminated-letter-d_371011.htm, public domain

Figure 41.1, Fixation Order, courtesy of Eye Track Shop

1. EyeTrackShop, "About Us," http://eyetrackshop.com/about-us.
2. Liz Gannes, "Eye Tracking Study Shows Users Perceive Google+ and Facebook Virtually Identically," *Wall Street Journal,* All Things D, August 11, 2011.

Chapter 42

1. "The Power of Visual Communication," Hewlett-Packard, July 2004, www.hp.com/large/ipg/assets/bus-solutions/power-of-visual-communication.pdf.
2. Christopher F. Chabris and Stephen M. Kosslyn, "Representational Correspondence as a Basic Principle of Diagram Design," Harvard University.
3. "Hans Rosling: Global Health Expert; Data Visionary," Speakers, TED Conference, www.ted.com/speakers/hans_rosling.html.
4. "Hans Rosling's 200 Countries, 200 Years, 4 Minutes—The Joy of Stats—BBC Four," YouTube, www.youtube.com/watch?v=jbkSRLYSojo
5. Natasha Singer, "When the Data Struts Its Stuff," *New York Times,* April 2, 2011.
6. Gapminder, www.gapminder.org/.
7. Gapminder, "Gapminder Desktop," www.gapminder.org/desktop/.

Chapter 43

1. visual.ly, FAQs, http://visual.ly/about/faq

Chapter 44

1. Deborah Landau, "The Best Readers Are Merciless Friends," *Wall Street Journal,* March 30, 2012.

Chapter 46

1. The 2011 Classic, TechAmerica, www.techamerica.org/classic.

Chapter 47

1. Stephen Sondheim, *Finishing the Hat: Collected Lyrics (1954-1981) with Attendant Comments, Principles, Heresies, Grudges, Whines and Anecdotes* (Knopf 2010), p. 7.
2. Paul Simon, "Isn't It Rich," *New York Times,* Sunday Book Review, October 27, 2010.
3. Nat Hentoff, "Protector of the Soulful Melody," *Wall Street Journal,* December 7, 2010.

Chapter 48

1. Dana Goodyear, "Hollywood Shadows," *The New Yorker,* March 21, 2011.
2. Stress Management Health Center, "Guided Imagery – Topic Overview," WebMD, www.webmd.com/balance/stress-management/tc/guided-imagery-topic-overview.
3. Elizabeth Quinn, "Visualization Techniques and Sports Performance," About.com, Sports Medicine, http://sportsmedicine.about.com/cs/sport_psych/a/aa091700a.htm.
4. Dana Goodyear, *op. cit.*
5. Dana Goodyear, *op. cit.*
6. W. Timothy Gallwey, *The Inner Game of Tennis* (Random House, 1974), p. 104.
7. Joel Stein, "The Shadow Knows," *Time Magazine,* June 4, 2012.

Chapter 50

Theme image, Taylor Mali, courtesy of Taylor Mali, who is a four-time National Poetry Slam champion. He is also the author of two collections of poetry and a book of essays, *What Teachers Make: In Praise of the Greatest Job in the World.*

1. *The Holy Bible,* King James Version, 1 Corinthians 13:11.
2. "Up-talking," Urban Dictionary, www.urbandictionary.com/define.php?term=Up-talking.
3. Taylor Mali, "Totally like whatever, you know?," Taylor Mali Poems Online, http://taylormali.com/poems-online/totally-like-whatever-you-know/.
4. Ronnie Bruce, "Typography," Vimeo.com, http://vimeo.com/3829682?ab.

Chapter 51

1. Matt Ridley, "Hands and Faces Spoke Long Before Our Tongues," *Wall Street Journal,* February 26, 2011.

Chapter 52

Figure 52.1, Ronald Reagan, GE Theatre, reproduced by permission of miSci

1. *Ronald Reagan: The Great Communicator*, Mpi Home Video, Release date June 15, 2004
2. Ronald Reagan, "A Time for Choosing," YouTube.com, www.youtube.com/watch?v=qXBswFfh6AY.
3. Irving Berlin, "Doin' What Comes Natur'lly," ST Lyrics, www.stlyrics.com/lyrics/anniegetyourgun/doinwhatcomesnaturlly.htm.

Chapter 54

1. Tony Perrottet, "Why Writers Belong Behind Bars," *New York Times,* Sunday Book Review, July 22, 2011.
2. Rolf Dobelli, "Avoid News," Dobelli.com, 2010, http://dobelli.com/wp-content/uploads/2010/08/Avoid_News_Part1_TEXT.pdf.
3. Freedom App, http://macfreedom.com/.

Chapter 55

1. Bruce Iliff, "Scuba Diving Anxiety and Panic," Water Sports @ Suite 101, February 2, 2008, http://suite101.com/article/scuba-diving-anxiety-and-panic-a43475.

Chapter 56

Theme image, Frank Sinatra recording session, courtesy of M. Garrett/Archive Photos/Getty Images

1. Microsoft Lync, Microsoft.com, http://lync.microsoft.com/en-us/Pages/unified-communications.aspx.

Chapter 57

Theme image, Nixon-Kennedy 1960 televised debate, public domain

Figure 57.1, The Rise and Fall of Rick Perry, reproduced by permission of Real Clear Politics, www.realclearpolitics.com

1. *2012 Election Central*, 2012 Primary Debate Schedule, http://www.2012presidentialelectionnews.com/2012-debate-schedule/2011-2012-primary-debate-schedule/.
2. Transcript: Fox News-Google GOP Debate, Fox News.com, September 22, 2011.

3. *Real Clear Politics*, 2012 Republication Presidential Nomination, http://www.realclearpolitics.com/epolls/2012/president/us/republican_presidential_nomination-1452.html
4. CNBC, "Your Money Your Vote" Republican Presidential Debate, *New York Times,* November 10, 2011.
5. Jon Meacham, "A Sense of Who They Are," *New York Times,* October 28, 2011.

Chapter 58

1. David Wiegand, "'Tonight – Four Decades of *Tonight Show*': Review," *San Francisco Chronicle,* December 22, 2010.
2. Dorothy Rabinowitz, "Johnny, We Hardly Knew You," *Wall Street Journal,* May 10, 2012.
3. IMDb, "Biography for Johnny Carson," www.imdb.com/name/nm0001992/bio.

Chapter 59

1. Adam Lashinsky, *Inside Apple: How America's Most Admired—and Secretive—Company Really Works* (Business Plus, 2012), p. 130.
2. Official Gmail Blog, "New in Labs: Undo Send," http://gmailblog.blogspot.hk/2009/03/new-in-labs-undo-send.html#!/2009/03/new-in-labs-undo-send.html.
3. Frank Partnoy, *Wait: The Art and Science of Delay* (PublicAffairs, 2012).

Chapter 60

1. Verne G. Kopytoff, "Yahoo Chief Executive Faces Unhappy Shareholders," *New York Times*, "Bits," June 23, 2011.
2. Yahoo! 2011 Annual Shareholders Meeting, June 23, 2011, www.shareholder.com/visitors/event/build2/mediapresentation.cfm?companyid=YHOO&mediaid=48223&mediauserid=5359581&TID=1304991612:c49e3db7614b66ae3bbcccc d45b4d8a0&popupcheck=0&shexp=201106280115&shkey=56b7dac451e96150 8a48182c9007d993&player=1.
3. *Wall Street Journal,* "Heard on the Street: Overheard," June 24, 2011.
4. Verne G. Kopytoff, "Yahoo Chief Executive Faces Unhappy Shareholders," *New York Times*, "Bits," June 23, 2011.
5. Don Reisinger, "Yahoo Annual Meeting Marked By What Wasn't Said," CNET.com, June 23, 2011.
6. *Ibid.*
7. David Streitfield, "Blunt E-mail Raises Issues Over Firing at Yahoo," *New York Times,* September 7, 2011.

Chapter 61

1. Scott Adams, *Dilbert,* September 25, 2010, www.dilbert.com/strips/comic/2010-09-25/.

Chapter 63

1. Adam Bryant, "Sit Down, Please, and Tell Me How Smart You Are," *New York Times,* September 18, 2010.
2. Adam Bryant, "Does Your Team Have the Four Essential Types," *New York Times,* October 2, 2010.

Chapter 64

1. Tim Weiner, "Robert S. McNamara, Architect of a Futile War, Dies at 93," *New York Times,* July 6, 2009.
2. Errol Morris, *The Fog of War:* Transcript, Errol Morris.com, http://errolmorris.com/film/fow_transcript.html.
3. CNN Transcripts, Western Republican Presidential Debate, October 18, 2011.
4. Lynn Sweet, "CNN Arizona Feb. 22, 2012 Debate Transcript," *Chicago Sun Times,* February 23, 2012.

Chapter 65

1. "Parolee Arrested Trying to Break Back into Sacramento Prison," *Los Angeles Times,* August 12, 2011.

Chapter 66

1. Dennis Lim, "A Generation Finds Its Mumble," *New York Times*, August 19, 2007.
2. George Bernard Shaw, *Candida,* (Penguin, 1950).
3. Alan Jay Lerner, *My Fair Lady*, book and lyrics (Coward-McCann, 1956), music by Frederick Loewe.

Chapter 73

1. Peter Funt, "The Lost Art of the Live Interview," *The Wall Street Journal*, April 9, 2012, page A13,.
2. Aspen Ideas Festival, "Charlie Rose: The Interview," http://www.aspenideas.org/session/charlie-rose-interview.

Chapter 74

Figure 74.1, courtesy of NetRoadshow, www.netroadshow.com

Chapter 75

Theme image, Cicero, http://upload.wikimedia.org/wikipedia/commons/1/18/Marcus_Tullius_Cicero.jpg, public domain

1. Moses Hadas ed., *The Basic Works of Cicero,* (Random House, 1951), p. 178.

Acknowledgements

My sincere gratitude to Silicon Valley venture capitalists John Doerr of Kleiner Perkins Caufield & Byers and Vinod Khosla of Khosla Ventures for sharing their unique perspectives on the mission critical importance of presentations, and for sharing their own best presentation practices. I am also grateful to John and Vinod for their continuing support of my coaching services.

Thanks, too, to the Power Presentations program participants who shared their own presentation experiences: Karen Wespi of Maxim Integrated, Peter Santos of Audience, Eric Benhamou of 3Com, Guillaume Estegassy of Microsoft, Steve Ahlbom of Artitudes Design, Noland Granberry of Silicon Image, Xavier Martin of Alcatel Lucent, Karyn Scott of Cisco, and Will Flash of Microsoft.

To the members of the Power Presentations team who support the programs that inspire the concepts and case studies in this book: Nichole Nears, Elenita De Lucca, Pearl Cheung, Rich Hall, Jim Welch, Bill Immerman, and Jon Bromberg.

To the publishing team from Pearson: Jeanne Glasser Levine, Megan Graue, Anne Goebel, and Geneil Breeze; to Warren Drabek from ExpressPermissions; to Jim Levine and Kerry Sparks from the Levine Greenberg Literary Agency; and an extra nod to Nichole Nears for managing the project, permissions, and graphics on our end.

To my blogosphere editors: Fred Allen @ Forbes, Geetesh Bajaj @ indezine.com, and Justin Fox @ *Harvard Business Review*.

To Robert Totman of NetRoadshow for his counsel on this exciting new form of communication.

For their story suggestions, thanks to Jim Fredricksen of Microsoft for "Winning and Losing the World Cup"; Dave Campbell of Microsoft for "Ready, Fire, Aim!"; Olivier Fontana of Microsoft (again!) also for “Ready, Fire, Aim!”; A. Gino Giglio for Taylor Mali in “Valley Girl Talk”; and Jeff Paine for Anchorperson/Weatherperson in “What do I do with my hands?”

To Norberto Vieyra for his lessons in Argentinian Spanish.

To the steady photographic hand of Jon Bromberg for his theme images.

To the image models, Elenita De Lucca, Master Brendan Lee Seals, and his swimming coach, Joon Young.

And to my lovely Lady Lucie for her love...

INDEX

Symbols

A

B

F

G

H

I

J

K

T

U

V

About the Author

Jerry Weissman is the world's number one corporate presentations coach. His private client list reads like a who's who of the world's best companies, including the top brass at Yahoo!, Intel, Intuit, Cisco, Microsoft, Netflix, Dolby Labs, eBay, HP, and many others.

Mr. Weissman founded Power Presentations, Ltd., in 1988. One of his earliest efforts was the Cisco Systems IPO road show. Following its successful launch, Don Valentine, of Sequoia Capital, and then chairman of Cisco's Board of Directors, attributed "at least two to three dollars" of the offering price to Mr. Weissman's coaching. That endorsement led to nearly 600 other IPO road show presentations that have raised hundreds of billions of dollars in the stock market. His techniques have helped another 600 public and privately held companies to develop and deliver their mission-critical presentations.

Mr. Weissman is also the author of four previous business books *Presenting to Win: The Art of Telling Your Story; The Power Presenter: Technique, Style, and Strategy from America's Top Speaking Coach; In the Line of Fire: How to Handle Tough Questions;* and *Presentations in Action: 80 Memorable Presentation Lessons from the Masters,* each of which has been translated into multiple languages worldwide.

As a prelude to Power Presentations, Mr. Weissman had an extensive career as a television producer/director, including a decade in the Public Affairs unit at WCBS-TV, the flagship station of the CBS Television Network in New York, and three years at the Network for continuing Medical Education.

In 1980, Pinnacle Books published Mr. Weissman's novel, *The Zodiac Killer*.

Mr. Weissman has a BA degree from New York University, and an MA in Speech and Drama from Stanford University.

FT Press
FINANCIAL TIMES